China-ASEAN Environment Outlook 1 (CAEO-1)

China-ASEAN Environmental Cooperation Center

# China-ASEAN Environment Outlook 1 (CAEO-1)

## Towards Green Development

CESP

Springer

China-ASEAN Environmental Cooperation
Center
Beijing
China

ISBN 978-981-10-6210-0 ISBN 978-981-10-6211-7 (eBook)
https://doi.org/10.1007/978-981-10-6211-7

Jointly published with China Environment Publishing Group Co., Ltd., Beijing, China

The print edition is not for sale in China Mainland. Customers from China Mainland please order the print book from: China Environment Publishing Group Co., Ltd.

Library of Congress Control Number: 2017949133



Printed on acid-free paper

This Springer imprint is published by Springer Nature
The registered company is Springer Nature Singapore Pte Ltd.
The registered company address is: 152 Beach Road, #21-01/04 Gateway East, Singapore 189721, Singapore

# A China-ASEAN Cooperation Fund Project

**Guided by:**
Ministry of Environmental Protection of People's Republic of China

**Supported by:**
ASEAN Secretariat

**Implemented by:**
China-ASEAN Environmental Cooperation Center

**Senior Advisory Group:**
Xu Qinghua
Li Haisheng
Guo Jing
Song Xiaozhi
Tu Ruihe
Tan Qingrui
Zhang Jieqing

**Working Group on China-ASEAN Environment Outlook:**
Chair: Zhou Guomei
Core Members:
Shi Feng
Jiang Hongqiang
Su Yang
Ge Chazhong
Huang Haifeng
Zhang Jinhua (United Nations Environment Programme)
Lailan Syaufina (Bogor Agricultural University of Indonesia)
Coordinator: Peng Ning

**Key Supporting Experts:**
Peng Ning
Wang Yufei

Cheng Cuiyun
Wu Huanan
Lan Yan
Li Xia
Chen Gang
Xie Ran

**Peer Reviewers:**
Li Lailai (World Resources Institute)
Niramon Sutummakid (Thammasat University of Thailand)
Clarissa Arida (ASEAN Center for Biodiversity)
Sheila Vergara (ASEAN Center for Biodiversity)

**ASEAN Secretariat:**
Raman Letchumanan (Former Head of Environment Division of ASEAN Secretariat)
Natalia Derodofa
Rocky Pairunan

# Foreword

The China-ASEAN Environment Outlook (CAEO) is supported by the China-ASEAN Cooperation Fund. Focusing on "Towards Green Development," it aims to carry out a comprehensive analysis of the status and progress of China-ASEAN green development and environmental protection, and look into trends of green development within the framework of global and regional environmental governance. This research report is prepared under the guidance and support of Ministry of Foreign Affairs, Ministry of Finance, and Ministry of Environmental Protection of People's Republic of China, approved and instructed by ASEAN Secretariat, and implemented by China-ASEAN Environmental Cooperation Center of Ministry of Environmental Protection.

To serve China-ASEAN Environmental Cooperation Strategy and Action Plan, and maintain the continuity and unique expert perspective of the research, it is necessary to build the Outlook into a flagship research report series. Research teams will choose topics of mutual interest and continue preparing and releasing China-ASEAN Environment Outlook in the future. Accordingly, this report is the inaugural edition, China-ASEAN Environment Outlook 1 (CAEO 1), marking an important starting point and achievement of China-ASEAN joint research.

In recent years, China-ASEAN cooperation has focused on economic fields, and gradually expanded to other fields including politics, security, culture, environment, forming a multi-level, wide-ranging, and all-round pattern. At present, China is building a moderately prosperous society and ecological civilization, and ASEAN is also accelerating the development of "green ASEAN" and rapidly pushing forward the integration process. Both sides have entered an important stage of development. China-ASEAN cooperation will also enter a new period of development.

Chinese Premier Li Keqiang depicted a new blueprint of "diamond decade" for China-ASEAN cooperation on the Seventeenth China-ASEAN (10+1) Leaders' Meeting in 2014, and further proposed "to jointly plan the grand strategy for China-ASEAN relations, jointly upgrade China-ASEAN FTA, accelerate the construction of interconnection network, carefully develop new points in maritime cooperation, make efforts to ensure security in both traditional and non-traditional fields, and actively explore the new fields for humanities, environmental protection, science and technology cooperation". He in particular proposed "to start developing 'ASEAN-China Strategy on Environmental Cooperation (2016–2020)', and share the achievements of scientific and technological innovation and eco-construction, in order to significantly improve the livelihood in each country, greatly accelerate the poverty reduction process, and benefit the people of both sides."

As one of the priority areas of cooperation within the framework of China-ASEAN cooperation, environmental protection attracts high attention from both China and ASEAN Member States. In 2009, China and ASEAN adopted "China-ASEAN Strategy on Environmental Cooperation (2009–2015)", which determined the priority areas of cooperation. In 2010, the Chinese Government approved the establishment of China-ASEAN Environmental Cooperation Center. Subsequently, China and ASEAN have jointly developed and adopted "ASEAN-China Environmental Cooperation Action Plan (2011–2013)", "China-ASEAN Environmental Cooperation Action Plan (2014–2015)", and "ASEAN-China Strategy on Environmental Cooperation (2016–2020)" and identify the focus of cooperation covering policy exchange and dialogue, green envoy program, environmental industries and technical cooperation, joint research, etc.

Such global environmental and development progress as 2030 Agenda for Sustainable Development and Paris Agreement on climate change is moving forward, and the process of international environmental governance is also accelerating. It becomes a trend to strengthen regional environmental governance and cooperation and achieve sustainable development via green transition. As one of the world's most economically dynamic regions, in the process of launching and constructing China-ASEAN Free Trade Area (FTA), China and ASEAN have formed a basic consensus on the necessity to achieve regional sustainable development through vigorous green economic development and inclusive growth, and to achieve regional sustainable development. Undoubtedly, a series of environmental and resources challenges are emerging, including those brought by population expansion, resources consumption, urbanization acceleration etc.

According to China-ASEAN Environmental Cooperation Action Plan, China-ASEAN Environmental Cooperation Center put forward the project proposal for research and development of a China-ASEAN Environment Outlook in 2015. The application was approved by China-ASEAN Cooperation Fund. According to the project plan, China-ASEAN Environmental Cooperation Center and ASEAN Secretariat jointly organized experts from China and ASEAN Member States to prepare "China-ASEAN Environment Outlook", analyze and assess the current condition and future trend of environment and development in China and ASEAN, and propose research perspectives and expert opinions on how to promote sustainable development in the region.

With "Towards Green Development" as its theme, the first edition of China-ASEAN Environment Outlook (CAEO-1) consists of six chapters to examine the commonly concerned regional green development issues in China and ASEAN. Chapter 1 "Overview of Green Development" reviews the origination of green development and defines green development for the report; discusses green development practices of different countries in the context of international development; and tries to use "pressure-state-response" (PSR) model to build an evaluation framework for green development. Chapter 2 "China-ASEAN Green Development Status" discusses in detail the current situation and key environmental issues in China's and ASEAN's green development, and makes a comparative study. Chapter 3 "Drivers, Features and Opportunities for Regional Green Development" sticks to the relationship between environment and development as a clue to describe development features of typical countries in the region and then analyzes the challenges and opportunities faced by various countries in achieving green development. Chapter 4 "Policy Measures for Regional Green Development" first reviews green development and environment-related policies and measures in China and ASEAN Member States and then makes in-depth analysis of typical policy measures and the associated benefits and limitations. Chapter 5 "China-ASEAN Green Development Practices" examines selected case studies on green industrial development, resource use, and climate strategies, pollution prevention and control, as well as biodiversity for China and ASEAN Member States including inter alia Thailand, Singapore, Viet Nam, the Philippines, and Cambodia. It provides basic information of background, effects, and implications of these cases, with the hope to inspire readers. Chapter 6 "An Outlook into the Future: Towards Green Transition" presents the forecasts of overall trends of environment and development in China-ASEAN region based on foregoing chapters, and puts forward policy recommendations on promoting cooperation on regional green development. On this basis, the Summary Report for Decision Makers is developed, which argues that the global 2030 Agenda for Sustainable Development brings significant opportunities to China and ASEAN Member States to deepen green development cooperation that the ongoing pragmatic cooperation should be pushed forward, including policy dialogue and exchange, environmental industries and technologies, capacity building, and green development best practices, and that the connection between "Belt and Road" initiative and green ASEAN Community should be established to achieve regional sustainable development.

It is hoped that the joint research of "China-ASEAN Environment Outlook" will build a knowledge-sharing platform to further promote understanding and consensus in the field of environmental development in China and ASEAN; at the same time, it provides relevant information and policy advice for implementing China-ASEAN environmental cooperation strategies and action plans, and promoting sustainable development in China and ASEAN Member States. Moreover, it is expected to depict a beautiful blueprint for and make a positive contribution to enabling China and ASEAN to achieve regional sustainable development.

Guomei Zhou
Feng Shi
Ning Peng

[illegible]

[illegible] China and ASEAN [illegible] development.

[illegible]
Feng Shi
Ning Peng

# Acknowledgements

China-ASEAN Environment Outlook is funded by the China-ASEAN Cooperation Fund, and sincere gratitude goes to the Ministry of Foreign Affairs (MFA), Ministry of Finance (MOF), and Ministry of Environmental Protection (MEP) of People's Republic of China (PRC) and to the ASEAN Secretariat for their review and approval of the project under the Fund. Gratitude also extends to the Department of International Cooperation of Ministry of Environmental Protection of PRC and Environment Division of ASEAN Secretariat for their guidance; to the leaders and divisions for Policy Research, China-ASEAN Cooperation, Asia-Pacific Cooperation, General Affairs Management, and Technology Cooperation of China-ASEAN Environmental Cooperation Center (CAEC) of MEP for their support and advice for project implementation and research report; to Director General Li Haisheng and Deputy Director General Song Xiaozhi at the Department of International Cooperation of MEP for their great support and suggestion. Director General Guo Jing of China-ASEAN Environmental Cooperation Center has devoted many efforts to this project since its approval, offered valuable suggestions on project research and implementation, and proposed strategic, forward-looking recommendations during the research, which comprehensively enhances the level and depth of the report.

It should be noted that any views expressed in this publication are those of the authors and do not necessarily represent the views of any official bodies or institutions. The authors and their research teams shall take sole responsibilities for the consequences of this publication.

# Contents

# Abbreviations

| | |
|---|---|
| ABO | ASEAN Biodiversity Outlook |
| ABS | Access and Benefit Sharing |
| ACCI | ASEAN Climate Change Initiative |
| AEC | ASEAN Economic Community |
| AEGE | ASEAN Expert Group on the Environment |
| ASCC | ASEAN Socio-Cultural Community |
| ASEAN | Association of Southeast Asian Nations |
| ASEAN-WEN | ASEAN Wildlife Enforcement Network |
| AMME | ASEAN Ministerial Meeting on Environment |
| AMS | ASEAN Member States |
| APMS | ASEAN Peatland Management Strategy |
| APRIL | APRIL Group |
| APSC | ASEAN Political-Security Community |
| AWGNCB | ASEAN Working Group on Nature Conservation and Biodiversity |
| BOD | Biochemical Oxygen Demand |
| CDM | Clean Development Mechanism |
| COD | Chemical Oxygen Demand |
| CPI | Consumer Price Index |
| DEWATS | Decentralized sewage treatment system |
| EEA | European Economic Area |
| EF | Ecological Footprint |
| EGS | Environmental goods and services |
| ENRAP | Environmental & Natural Resources Accounting Project |
| EPI | Environmental Performance Index |
| ESI | Environmental Sustainability Index |
| FAO | Food and Agriculture Organization |
| FSC | Forest Stewardship Council |
| FYP | Five-Year Plan |
| GBI | Green Building Index |

| | |
|---|---|
| GDP | Gross Domestic Product |
| GEO-6 | Global Environment Outlook 6th Edition |
| GEP | Green Envoy Program |
| GHG | Greenhouse Gas |
| GMS | Greater Mekong Subregion Cooperation |
| GTI | Global Taxonomy Initiative |
| HCVF | High Conservation Value Forest |
| HDI | Human Developing Index |
| IAS | Invasive Alien Species |
| IEA | International Energy Agency |
| IMF | International Monetary Fund |
| IPCC | Intergovernmental Panel on Climate Change |
| IRP | International Resource Panel |
| ISO | International Organization for Standardization |
| IUCN | International Union for Conservation of Nature |
| LCCF | Low carbon city framework agreement and evaluation system |
| MBI | Market-based Instruments |
| MPAs | Marine Protected Areas |
| MRFs | Material Recovery Facility |
| MSW | Municipal Solid Waste |
| NBSAP | National Biodiversity Strategy and Action Plan |
| NEB | Net Environmental Benefit |
| $NO_x$ | Nitrogen Oxides |
| NSCB | National Statistical Coordination Board |
| NSS | National Space Society |
| OECD | Organization for Economic Co-operation and Development |
| PECSME | Promoting Energy Conservation in Small and Medium Scale Enterprises |
| PM | Particulate Matter |
| PSEEA | Philippine System of Integrated Environmental and Economic Accounting |
| PSR | Pressure-State-Response |
| RCEP | Regional Comprehensive Economic Partnership |
| RDFPs | Refuse Derived Fuel Power Plant |
| REPI | Resource Environmental Performance Index |
| RHAP | Regional Haze Action Plan |
| RSPO | Roundtable on sustainable palm oil initiative |
| SAN | Sustainable Agriculture Network |
| SBMSP | Science-Based Management Support project |
| SCG | Siam Cement Public Company Limited |
| SCP | Sustainable Consumption and Production |
| SDGs | Sustainable Development Goals |
| SEC | Singapore Environment Council |
| SEEA | System of Integrated Environmental and Economic Accounting |

| | |
|---|---|
| SERIEE | European System for the Collection of Economic Information on the Environment |
| SGLS | Singapore Green Labelling Scheme |
| $SO_2$ | Sulfur Dioxide |
| UNCED | United Nations Conference on Environment and Development |
| UNCLOS | United Nations Convention on the Law of the Sea |
| UNDP | United Nations Development Programme |
| UNEP | United Nations Environment Programme |
| UNESCAP | United Nations Economic and Social Commission for Asia and the Pacific |
| UNFCCC | United Nations Framework Convention on Climate Change |
| UMTNP | U Minh Thuong National Park |
| VOC | Volatile Organic Compounds |
| WCED | World Commission on Environment and Development |
| WDPA | World Database of Protected Areas |

# List of Figures

# List of Tables

# List of Boxes

# Summary Report for Decision Makers

In September 2015, heads of governments around the world adopted the 2030 Agenda for Sustainable Development at the UN Sustainable Development Summit. Serving as a new global development agenda succeeding the Millennium Development Goals (MDGs), the 2030 Agenda highlights 17 Sustainable Development Goals (SDGs), with 169 associate targets, further emphasizing an integrated approach and comprehensive solutions toward coordinated economic, environmental, and social development. In December 2015, nearly 200 UNFCCC Parties adopted the Paris Agreement, which provides new institutional framework and arrangements for global response to climate change after 2020 and underpins the shift of global economy toward curbing emissions and strengthening resilience to climate change. These two milestone events indicate that the international community has already reached consensus on sustainable development with well-shaped philosophy, theory and corresponding institutional arrangements, and is moving toward the implementation phase. It also signifies the emergence of a new trend for global environmental governance.

China and ASEAN Member States have been making their best efforts in experimenting and implementing policies and practices of green development according to their own conditions. In the context of implementing the 2030 Agenda and the Paris Agreement, great potentials and opportunities will be released for China and ASEAN to boost regional environmental protection and development.

**Overview of Green Development**

Since entering the twenty-first century, the world has been experiencing increasingly serious environmental pollution and resource exhaustion, which results in a number of crises relating to climate, biodiversity, water, food, energy, and various resources. These stimulate constant reflections on the prevailing economic development model featuring "high energy consumption, high pollution, and high ecological damage" and give rise to thinking and research on a new development model. With the growing public understanding of links between economic activities, resources, and the environment, and in response to the 2008 financial crisis, green development came into being, as an effective approach toward economic development, environmental protection, and social inclusion. China frames a development path with five concepts, i.e., innovation, coordination, green, opening and sharing, further contributing to the consolidated foundation of theory, institutional set-up and practice for green development.

**Basic Features of Green Development: Resource Conservation, High Energy Efficiency, and Environmentally Friendly**

This report defines green development as a model and institutional innovation based on traditional development, and a new development model and path in favor of resource conservation and environmental protection, established under the constraints of environmental capacity and resources carrying capacity.

With an overall aim to realize the coordination between economic development and resources and the environment, green development works to reduce the impact on resources and the environment while seeking economic growth. Green development will attain three objectives: (1) to address preferentially the problems of resources and environment facing mankind through energy conservation and emission reduction; (2) to improve resource efficiency and green competitiveness of industries by virtue of technological progress; and (3) to transform the economic development mode, adjust, and improve structure, and promote the green development of the economic system by means of green transition.

In essence, green development requires energy-saving and environment-friendly economic activities. It is directed at minimizing environmental load and impacts caused by economic activities through available and appropriate technologies and economic conditions, to finally control such load and impacts within the resource supply capacity and environmental self-purification capacity of ecosystems. It aims to ensure good ecosystem status and human health, and bring forth a virtuous cycle between socioeconomic activities and ecosystems. Specifically, green development focuses on “low consumption, low emission, and low pollution,” and “high benefit, high carbon sink, and high cycle.”

In the implementation process, green development shares the same goals with sustainable development. Green development enables the balanced development of economy, environment, and society through transformation of economic development models, reduction of resource consumption and pollution emissions, and promotion of social progress and fairness. Therefore, green development is an important means to tackle climate change and realize sustainable development.

**Construction of Indicator Systems and Methods to Evaluate Green Development**

The 2030 Agenda stresses the need to monitor and assess progress in implementing SDGs. In the same way, monitoring and evaluation are also conducive to advancing the process of green development.

Green development aims to coordinate economic growth and the carrying capacity of resource, environment, and ecosystems. Therefore, to evaluate how green the development could be and the degree of green transition, it should consider costs of consumption and pollution related to resources and environment caused by economic activities and explain stock and flows of resources (e.g., using

System of Integrated Environment and Economic Accounting, implementing Green National Accounting). The evaluation system should also fully recognize the economic value of natural resources, with consideration of economic transition, resource efficiency, social development, and human well-being.

Based on existing assessment systems for green development, this report suggests that the "pressure-state-response" (PSR) model would be adopted as a choice for building green development assessment system. The PSR model reflects mainly the causal relationship between human activities, environment, and society: human activities impose certain impacts on the environment; the environment performs self-regulation within its limits; and the society gives a response according to environmental status so as to maintain a healthy and stable environment (Fig. 1).

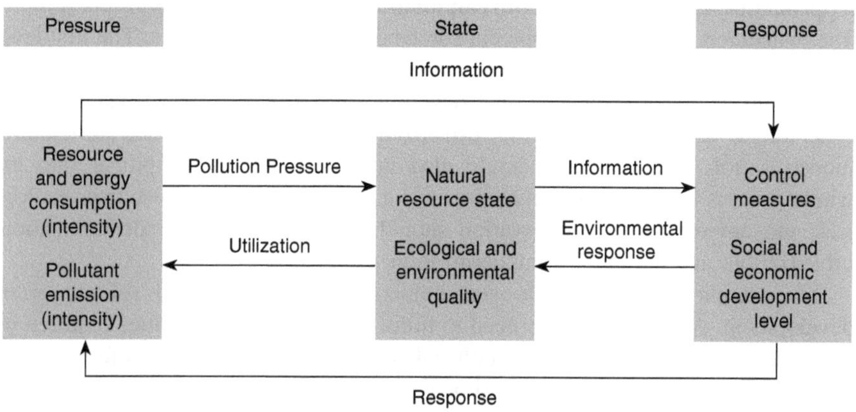


**Fig. 1** PSR Model

The assessment system for green development is based on the three modules of "Pressure-State-Response". Among them, the quality of environment and natural resources corresponds to the "State" part of the PSR model; response measures taken by human society to solve problems and social economic development levels correspond to the "Response" part of the PSR model; impacts imposed by human activities on the environment correspond to the "Pressure" part of the PSR model. Users can choose "localized" indicators that are specific to their own conditions under the PSR framework to gauge their progress in green development. For instance, the green economic development index could combine energy consumption, environmental quality, and economic growth to measure the development of green economy.

The central pillar of green development is green economy. The connotation and extension of green development are greater than those of green economy. Modularized indicator system can be established based on analysis of key elements that underpin green development, and then used for evaluation with inputs from scenario analysis of specific development situations. Core elements in the modules will include: economic and social development (including economic aggregate, growth rate, industrial structure related to primary, secondary and tertiary industries, levels of industrialization and urbanization, social welfare and security), intensity and efficiency of resource utilization, environmental quality (air, water, soil), energy efficiency and climate change, ecosystem and biodiversity conservation, sustainable production and consumption, and institutional arrangements (capital investment, technological innovation, regulations, policies, standards, etc.). These core elements are interconnected systems. The aggregate, speed and structure of economic development directly affect the quality of eco-environment; technological innovation and institutional arrangements are also core determinants of eco-environmental quality; ecosystem services play an important supporting role in economic and social development. This report discusses and compares the core elements of the above modules of green development.

**Approaches Towards Green Development**

The realization of green development can be divided into two levels. The first level deals with green development of industries. Green development should cover all levels and factors of socioeconomic development, i.e., green development should synchronize with industrialization, urbanization, informatization, and agricultural modernization. Governments should also integrate green development into the whole process of economic development, i.e., considerations of resource, ecological, and environmental conservation should be integrated into industrialization, urbanization, and agricultural modernization, etc.

The second level relates to stakeholders' responsibility and role in green development. All social players need to fulfill their due responsibilities required by green development to establish a multistakeholder governance system for boosting green development. Governments should play a guiding and regulating role, provide adequate systems to internalize eco-environmental externalities, and implement green management to form harmony between man and nature, and to coordinate economic, social, and environmental development. Enterprises are entities for practice and innovation for green development and should actively fulfill their social environmental responsibility. It is necessary to integrate green driving forces to fuel green development. Public awareness and practice of environmental protection, resource conservation and sustainable consumption should be encouraged.

- To transform economic development model, adjust and optimize industrial structure. It is important to fundamentally transform national economic growth model and increase the share of green development in national development through adjustment, optimization and upgrading of industrial structure. Innovations in business models, goods, and services will leverage a greener economy, and promote green economic development.
- To advance industrial transformation, upgrading and green manufacturing. It is necessary to promote the transformation of industrialization process, with particular emphasis on the development of low-pollution and high-efficiency technologies. Such transition will be enabled by the modernization of manufacturing industry, development of artificial intelligence and the Internet+ technology. At the same time, great efforts will be needed to reduce energy and

resource consumption and strengthen pollution management as optimized resource and energy mix and improved energy efficiency are key to green development.

- To promote the process of green urbanization. To promote green urban development, more attention should be paid to enhancing intrinsic qualities such as urban culture, characteristic industries, livable environment, and social services. It is necessary to give full play to the services and leading effect of the tertiary industry, focus on solutions to traffic congestion, air pollution, water and food security, and build a livable, green, inclusive, and flexible urbanization model.
- To implement agricultural modernization with high-efficiency and low-environmental impact. Through technical innovation, existing dominant industrial and agricultural technologies will be further upgraded to reduce environmental pollution, improve efficiency, and promote precision agriculture with low environmental impact. In addition, great efforts are needed to boost development of circular eco-agriculture.
- To accelerate independent innovation and application of green technology using the market mechanism. Measures should be taken to (1) encourage innovation, research, development, and application of green technologies, (2) make full use of market mechanism to facilitate application/demonstration pilot projects on green scientific and technological development, and (3) further accelerate the development and industrialization of environmentally friendly technologies (including technologies for transforming traditional industries, emerging industries and environmental industries).
- To boost supply-side reform with polices and establish a package of economic policies, laws and regulations in favor of green development. Government can employ integrated measures of market, laws and policies to force producers to internalize environmental externalities that they imposed on society and encourage producers to make innovation in green technology and perform green transition. Moreover, governments should accelerate the ecological innovation of traditional industrial standards, and establish a green standard system covering all fields.
- To foster public awareness of and actions for sustainability, enable multi-stakeholder governance, and create stronger social forces to boost green development. A variety of multilevel publicity and education programs are needed to popularize green development knowledge and make the general public aware of the significance and necessity of low-carbon life. Efforts should be made to guide consumers through proper policy measures to practice low-carbon life and sustainable, low-carbon, green consumption.

**Green Development Practice and Environmental Protection in China and ASEAN**

The Asia-Pacific region, especially the ASEAN region, is one of the most dynamic regions in world economy. Rapid industrialization and urbanization has promoted the integration of regional economy. Although China and ASEAN Member States are not identical in economic and political systems, development stages, resource endowment, and geographic environment, they are all faced with environmental pollution, excessive and inefficient use of resources, and other problems brought by rapid economic development; they also have to solve the problems of booming population growth, a widening gap between the rich and the poor, and other social problems. Green development has provided a strategic option for China and ASEAN Member States in the future to solve the above problems.

Although green development is becoming a general trend, it should be noted that there is no "generally applicable" green economic transition model. China and ASEAN Member States are mostly developing countries, and they are different in history, politics, and specific conditions, as well as levels of economic development. In order to truly realize green development, countries must determine suitable green development paths after considering their own development stages, and identifying suitable development goals and constraints.

**Green Development and Environmental Protection in China**

China is an early player of green development. For the relationship between the environment and economic growth, China's leadership and policy researchers have put forward the concepts relating to ecological civilization and green development over more than a decade, ranging from Scientific Outlook on Development to "resource-efficient and environmentally friendly society", from ecological civilization integrating into "Five-in-One" layout to the five concepts of "innovation, coordination, green, open, and sharing". Green development is playing a leading role in China's new normal of economic development.

China has witnessed over three decades of high-speed growth, and become the largest exporter and manufacturer and the second largest economy. While its economy develops rapidly, China's poverty rate has also dropped from 65% to less than 10%, with more than 500 million people lifted out of poverty and the MDGs achieved. At present, China has entered the New Normal state and maintains the economic growth driven by the secondary industry, while speeding up structural adjustment. Strategic emerging industries, including those of energy conservation and environmental protection, new-generation information technology, biology, high-end equipment manufacturing, new energy, new material, and new energy automobile, are showing promising momentum, with expanding industrial scale.

China's energy mix is being optimized, and the intensity of carbon emissions has decreased significantly. Coal has always accounted for a big share in the energy mix. As the relatively clean natural gas and other energy resources (hydropower, nuclear power, wind power, etc.) are applied gradually, China's energy consumption per unit of GDP is in decline. According to Trends in Global $CO_2$ Emissions 2015 Report, China's carbon emission is rising, yet the emission growth rate slows down. Compared with other countries, China witnesses a significant and faster decrease in carbon emission intensity. With regard to resource use, China's national land development has kept rising, with uneven fluctuations on the annual increment. There is an urgent need to improve resource efficiency in land and water use.

Environmental protection has received overwhelming attention, and positive progress has been made. Since the 12th Five-Year Plan (FYP) Period (2011–2015), legislative and institutional frameworks have been improved, with the promulgation of the most stringent Environmental Protection Law, as well as Action Plan for Air Pollution Prevention and Control, Action Plan for Water Pollution Prevention and Control, and Action Plan for Soil Pollution Prevention and Control. Pollution control actions have generated fruitful results, with gradually decreased total emissions of COD, ammonia nitrogen, and heavy metals, as well as decreased $SO_2$ emission intensity. However, composite pollutants such as $NOx$, ozone, and particulate matters have been on the rise. Biodiversity conservation eyed positive progress. The eco-environment quality is improving, and the contribution of environmental protection to economic development is increasing.

In terms of pollution control, China's urban sewage treatment capacity has increased from 125 million tons per day in 2010 to 182 million tons per day in 2015, becoming one of the countries with the strongest sewage treatment capacity. China has promoted ultra-low emissions in coal-fired power units and established the world's largest clean and efficient coal power system,[1] with revolutionary progress achieved. Sulfur dioxide and nitrogen oxide emissions from thermal power industry have decreased by 47% and 50%, respectively, and the COD discharge and ammonia nitrogen emissions per unit of industrial value-added have decreased by 42% and 48%, respectively.

In terms of ecological protection and biodiversity conservation, China accelerated the ecological redlining to optimize the spatial layout of development, and proposed the binding requirements of green development from the perspective of carrying capacity of resources and environment. As of 2015, China has built 2,740 nature reserves, accounting for 14.8% of the total land area, higher than the world's average of 12.7%.

Up to now, China has built up the environmental labeling system for products, the environmental labeling system for manufacturing and building products, and the energy efficiency labeling system. It is now actively promoting sustainable production, consumption, and lifestyle.

Through above measures, China's COD, ammonia nitrogen, sulfur dioxide, and nitrogen oxide emissions have decreased by 12.9, 13, 18, and 18.6% by 2015, respectively, compared with the 2010 level, and emission reduction targets in the 12th FYP have been met. The most obvious change is the positive trend in environmental quality. Air quality as a whole in China is getting better. In 2015, the average concentration of $PM_{2.5}$ in 74 key cities has fallen by 14.1% over 2014 and by 23.6% over 2013.[2]

China is also actively promoting regional cooperation on environmental protection, in order to make positive contributions to solving global environmental problems. For example, under the framework of Vienna Convention for the Protection of the Ozone Layer and the Montreal Protocol on Substances that

---

[1]Huang Runqiu: A New Blueprint For Environmental Protection, China Economic Report, 4th issue in 2016.

[2]See footnote 1.

Deplete the Ozone Layer, the elimination of ozone depleting substance (ODS) in China accounts for more than 50% of the total elimination in all developing countries, making the world's largest contribution to the protection of ozone layer. China actively carries out South–South cooperation with developing countries and promotes environmental cooperation under such frameworks as ASEAN and Shanghai Cooperation Organization, all of which generate positive results. Especially, environmental cooperation between China and ASEAN has successfully extended to new forms such as developing cooperation strategy and action plans, policy dialogues and exchanges, technical cooperation on environmental industries, personnel exchanges and capacity building, and joint policy research. China and ASEAN have created a model for South–South cooperation.

**Green Development and Environmental Protection in ASEAN**

Over the past decades, ASEAN Member States have made significant stride in economic growth, with fast growing population and accelerated urbanization. ASEAN has become an area with strong economic vitality. However, the economic scale differs among member states. In 2013, Singapore, Brunei Darussalam and Malaysia were in the development leaders, with the per capita GDP being 10,000 US dollars or more, which was 4,000–8,000 US dollars for other member states. In recent years, ASEAN Member States have experienced steady economic growth.

On the whole, fossil fuel continues to dominate the energy mix of ASEAN. Gross energy use keeps growing along with the constant increase of population and GDP, and its growth stays consistent with the increase of per capita income and the middle class emerging in the region. From 1990 to 2007, the gross energy use of ASEAN doubled. In the meanwhile, the use of coal and natural gas increased and the fossil fuel-to-gross energy demand rate rose, while the ratio of petroleum remained basically unchanged. The use of non-fossil fuel witnessed a decrease because commercial fuel substituted traditional energy and the proportion of biomass fuel (generally including forest and agricultural residue) dropped. To foster a shift in economy toward sustainable industrial development, ASEAN Member States have launched cleaner production plans in manufacturing successively, such as Malaysia's Roadmap for Cleaner Production, Viet Nam's National Clean Production Action Plan, Thailand's National Clean Production Development Plan, the Philippines Environmental Partnership Plan, the Singapore Green Label Scheme (SGLS), and so on.

Although ASEAN Member States are not sources of significant greenhouse gas emissions, they are very sensitive to climate risks. They have taken various actions to address climate change, with special focus on reducing natural disasters. Measures include encouraging afforestation and/or reforestation to mitigate carbon emissions from deforestation and/or forest degradation, improving forest management, building resilience to climate change, and increasing energy efficiency in power production, construction, industries, and transportation.

Many ASEAN Member States have initiated actions to incorporate climate change adaptation measures into development plans to strengthen resilience. Such regional level efforts as ASEAN Climate Change Initiative (ACCI) are effective. It is worth noting that the current economic crisis is an opportunity for ASEAN Member States to switch to low-carbon economy and adapt to climate change. It is also an opportunity to upgrade "green investment" plans into fiscal stimulus plans that incorporate adaptation and mitigation measures with actions to strengthen economy, create jobs, and reduce poverty.

Air pollution is among the most severe environmental problems for ASEAN Member States. As a result of increasing fossil and biomass energy use, booming population, and accelerated urbanization, the air quality of some ASEAN Member States is worrisome. Industrial, agricultural, and transportation sectors are the major sources of air pollution. Most ASEAN Member States have set up monitoring networks for air quality. Cooperation agreements on transboundary haze and air pollution control were reached, such as ASEAN Cooperation Agreement on Transboundary Haze Pollution.

The eco-label plan and environment-friendly product plan have been implemented in many ASEAN Member States to encourage sustainable production. As demand for green products is growing, green products and corresponding industries still have great potential for growth.

ASEAN also makes efforts in various respects to fulfill its promise to promote green development at regional level and has promulgated a great many declarations and proposals relating to environment. Major developing economies of ASEAN and other Asian countries jointly promised to establish a framework of policies, monitoring and regulations so as to promote the low-carbon and efficient transition of industry. ASEAN Member States are devoted to regional green development for an ASEAN Community through coordination, cooperation, and communications.

**Comparative Results on China-ASEAN Green Development and Environmental Protection**

The examination and comparison of green development in China and ASEAN reveal following characteristics of the two side:

- Economic development is relatively decoupled from resources and environment, but the challenge is still grim. After years of efforts, China's economic development is showing a trend of relative decoupling from resource consumption and environmental degradation, but no strong tendency of decoupling is observed in most ASEAN Member States. It indicates that economic development is still attained at the expense of huge energy consumption and natural environment deterioration in most ASEAN Member States.
- There are imbalances in terms of green development progress and priorities, and it is a difficult challenge to balance development and protection. China and ASEAN make efforts from various aspects to fulfill their commitment to green development, but have great differences in terms of economic development level, natural resource endowment as well as the progress and priorities of green development.

It is an arduous and multiaspect challenge for both China and ASEAN to manage the natural resources through maintaining a balance between protection and exploitation.

- Air pollution and water pollution are common challenges, and the protection of public health is a priority area and key issue. Similar to China, ASEAN Member States are generally faced with severe water and air pollution due to the increasing population, speeding urbanization and influence from factors such as industrial production activities. But they have made progress in pollution control.
- The mix of energy consumption is still dominated by fossil fuels, but is showing promising trend in structural change. The gross energy consumption keeps growing along with the unceasing expansion of population and GDP and energy mix is adjusted step by step. The difference is that ASEAN energy mix is based on petroleum consumption while China's coal-focused structure can be hardly changed in a short term.
- Protection of forest resources is a priority area for ecosystem protection. ASEAN is among the regions with the densest and most extensive forest resources, with a regional average coverage rate of 47%, much higher than that of China, 21.63%. However, the forest area is decreasing in several ASEAN Member States, along with the increasing population, and the growth of agricultural production, logging, and mining.
- The rapid progress of industrialization and urbanization has produced soaring amount of waste, requiring a sustainable solution. The increasing number of waste emissions has become a major challenge for green development. Therefore, waste management has become an important issue for China and ASEAN. Urban areas witness a big rate of wastes generation, which is aggravated further by other environmental problems arising from there.

**Effective Measures and Practices in Green Development**

In pursuing green development, China and ASEAN Member States have developed various plans and policies specific to their own circumstances and formed a portfolio of effective measures and practices. In this report, some typical green development cases in China and ASEAN Member States are chosen to reflect some good grassroots practices of green development. These good practices prove that "all theory is gray, but forever green is the tree of life" (Johann Wolfgang von Goethe). China and ASEAN Member States can further share and communicate these good practices of green development and environmental protection with each other. These good practices of green development unexceptionally share the following common features.

(1) **Effective environmental laws and policies**

China and most ASEAN Member States have put in place a relatively complete environmental governance system and legal system. Corresponding policies are working well in protecting environment. Currently, the command-and-control policy plays a dominant role in the region, while the environment-economic policy and voluntary environmental policy are under progressive development, with growing popularity among decision makers and great potential to be widely used.

(2) **Incentives and measures in favor of green economic growth**
China and ASEAN Member States implemented incentives and measures in favor of circular economy, green industries, cleaner production, comprehensive utilization of resources, and environmental industries. These incentives and measures include: establishment of a sound environmental property right system, incentives to improve resource productivity, such environment-related economic instruments as green credit, pollution liability insurance, and green investment, all of which support institutional arranges for green development. In addition, awards and certifications are also proven effective.

(3) **Technological innovation to drive green development**
Technological innovation has significant implications to industrial upgrade and business expansion, and is deemed as a primary driving force for green transition of economic development. Governments should continue encouraging green technological innovation and provide strong legal protection for green inventions by improving intellectual property system for green innovation.
Multistakeholder participation also plays an important role in implementing the above-mentioned policy measures. A successful practice is probably a result of governments' guidance, enterprises' participation, public's acceptance and support, assistance from NGOs, and coordinated actions and cooperation from different regions/countries.

**China-ASEAN Cooperation Prospects in Environment and Development: Strategic Choice of Green Development at the Crossroads of Development**

China is connected to ASEAN by lands and seas. Over the past years since China and ASEAN opened dialogue in 1991, the two sides have increasingly supported each other in development. In light of global economic growth, Asia and the Pacific is viewed as the region with the fastest growth rate in the world. ASEAN and China are likely to become important growth engines in Asia-Pacific. ASEAN has launched negotiations on the Regional Comprehensive Economic Partnership (RCEP), making crucial contributions to the integrated development in the region. Initiatives such as the twenty-first Century Maritime Silk Road and the Asia Infrastructure Investment Bank proposed by China will also reinforce regional connectivity and facilitate the building of a community of common destiny between China and ASEAN.

**Economic and Social Development**

(1) **China: Making the strategic choice of green transition at the crossroads**
The future 5–10 years will become a critical period for China to achieve its mid- and long-term socioeconomic development goals and a window of opportunity for it to achieve its green development. China is now standing at the crossroads of development. At this crossroads and during this key opportunity period toward the sustainable development, China encounters a series of

challenges. It will achieve the long-term sustainable development and build a beautiful China after making the strategic choice of green transition.

- To avoid the Middle-Income Trap[3] in economic and social development. On the development path, Latin American countries have provided a profound lesson from their failure in escaping the middle-income trap. According to World Bank standards, China is moving toward an upper middle-income country. It has to confront a series of challenges and deal with the relations of the economy, society, and environment so as to escape the middle-income trap. While stepping into the new normal state, China's economic growth will switch from high speed to moderately high speed; the industrial structure starts to turn from the secondary industry-dominated pattern to the tertiary industry-dominated pattern; and the demand structure changes from investment-dominated pattern to consumption-dominated pattern. China will face a fast growing aging population, a shrinking working-age population, and deal with the disappearing population dividend. Strengthened efforts are needed to adjust the population policy and social security system.
- To modernize national governance system by enhancing governments' green leadership and avoiding Tacitus Trap.[4] To enhance the environmental governance capability of government and society as a whole requires an active, comprehensive, and transparent disclosure of the environmental information, full use of big data and information technology, and establishment of the environmental information sharing and support platform. When dealing with environmental public affairs, the laws and regulations and the procedural fairness and justice should be strictly abided by to avoid the recurring Not-In-My-Back-Yard (NIMBY) events. In addition, governments should be aware that the public's increasing demand for high-quality life and diversity of social interests and aspirations will raise the share of consumption in GDP. Green and sustainable consumption should become a driving force for the social green development.

[3]Latin American Middle-Income Trap. When the Gross Domestic Product (GDP) per capita of a nation is between 1,000 and 3,000 US dollars, it is usually the time when the growth and problems, development, and conflicts are intertwined and the time when social structure undergoes profound changes and paradoxes can be easily intensified. Since 1980s, some countries in Latin America experienced an increasing economic growth and their GDP per capita reached 2,000 US dollars at the beginning of the twenty-first century. During this period, unemployment rate kept increasing; the gap between the rich and the poor exacerbated; various social conflicts exhibited and then intensified; the social upheaval and unrest struck; and the mass uprising occurred one after another.

[4]The term "Tacitus Trap" comes from the historian Tacitus (AD 55?–117) during the Ancient Rome Times. It roughly means when a ruler once becomes unpopular, all his acts, be they good or bad, tell against him. Such a remarkable opinion later became one law in Western politics, that is, "Tacitus Trap".

- To promote the regional coordination and rebalance development strategy in the national regional development. "Silk Road Economic Belt" and "21st Century Maritime Silk Road" (the "Belt and Road" initiative), Jing-Jin-Ji (Beijing–Tianjin–Hebei) Economic Integration, and Yangtze River Economic Belt will not only boost regional economic development in China but also reshape the regional economic geographic pattern and even the global environmental and developmental pattern. Urbanization will be sustainably and rapidly developed and the overall layout and inherent quality of urbanization will be further optimized and improved.
- To avoid Thucydides's Trap[5] while dealing with the geopolitics, international relations and big power relations. China launches enormous efforts to build a new-type relationship with the USA, and carry out the South–South cooperation, with developing countries, in response to the thematic requirement of peace and development in this age. What is urgent is to deal with the common challenges facing human beings, which mainly include climate change, terrorism, transnational crime, pandemics, natural disasters, proliferation of weapons of mass destruction, and other challenges. Human survival heavily depends on cooperation and concerted efforts.

In November 2014, China and the USA released the first joint statement on climate change. In September 2015, President Xi Jinping and President Barack Obama released the second joint statement on climate change during Xi's state visit to the USA. On March 31, 2016, both parties released the third joint statement and agreed to sign Paris Agreement on April 22. These three joint statements make substantial contribution to global response to climate change. Taking joint actions to address common challenges is becoming an increasingly important area for Sino-US cooperation. Enormous potential and opportunities exist in the bilateral cooperation, such as Sino-USA cooperation and China-ASEAN cooperation, in the fields of the economy and trade, science and technology, education, humanity, environmental protection, and so on.

(2) **ASEAN: Building an economic, social, and environmental community**
ASEAN will step into the era of ASEAN Community, with established ASEAN Political-Security Community (APSC), ASEAN Economic Community (AEC), and ASEAN Socio-Cultural Community (ASCC).

- It is reckoned that the average GDP of ASEAN will reach 5.4% by 2030, with the projected per capita income of around 12,000 US dollars.[6] The implementation of AEC could create an annual economic value between

[5]Thucydides's Trap refers to the phenomenon that an emerging power must challenge the old one who will definitely respond to such a threat, thus making the war inevitable. It stems from Thucydides (BC 460/465-400), the well-known historian in Ancient Greece, who stated that the competition between an emerging power and the old one will mostly end with the war.

[6]ADB. ASEAN, PRC and India: the Great Transformation. 2014.

280 billion US dollars to 615 billion US dollars for ASEAN by 2030, which would be equivalent to 5%–12% of ASEAN GDP in 2030.[7]

- It is projected that the full implementation of AEC will generate by 2025 an additional 14 million jobs in six ASEAN economies (Cambodia, Indonesia, the Lao PDR, the Philippines, Thailand, and Viet Nam), speed up the pace of structural change, and drive up productivity.[8]
- The living standards for ASEAN population would rise substantially, with enhanced middle-income groups. Against the standard of 10 US dollars to 100 US dollars per capita per day, around 64% of ASEAN population will be among the middle-income group by 2030.
- Urban population is supposed to account for 56% of the total population in ASEAN by 2030. Advantages in population trend, sustained urbanization, soaring productivity, and improved technologies will become major forces to fuel ASEAN's development.

**Regional Environment and Development Trends**

It is predicted in the GEO-6 Regional Environmental Outlook of Asia and Pacific that the Asia-Pacific population, lifestyle, and the access to basic environmental services will undergo rapid changes in the future.[9] Based on China-ASEAN environmental and developmental characteristics, challenges and trends of this region can be summarized as follows.

Firstly, the energy efficiency waits to be further improved as the energy consumption increases. With the expansion of the economic scale, industrialization leads to accelerated development of urbanization. Resource consumption will soar, but resource and energy efficiency will have little improvement, causing widespread environmental deterioration, degradation of ecosystem services, and excessive waste, creating additional economic burden.

Secondly, there is an increasing impact of the natural disasters and extreme events that are caused by the climate change. The urbanization and urban immigrants gradually weaken the regional capability of confronting climate change and disasters, which will further lead to economic losses, offset development gains, increase the poverty and inequity, and pose threats on water supply and food security.

Thirdly, the health risks associated with environmental pollution and the public's attention to these risks show a trend of increase. The emergence of contagious diseases related to various pollution (e.g., air pollution, hazardous chemicals and heavy metal pollution, and water pollution) leads to the public's increasing concern about the health risks associated with the environmental pollution. Such public concerns even become obstacles to site selection for the environmental pollution control infrastructure, resulting in emergence of NIMBY events.

Fourthly, there is an increasingly widening gap between the environmental regulations and policies and their implementation, resulting in the low efficiency

[7]http://www.adb.org/sites/default/files/publication/160067/ewp-432.pdf.

[8]See footnote 7.

[9]UNEP, Regional Environmental Outlook of Asia and Pacific, 2016.

and inexplicit effects in policy implementation. In response to existing situation, governments have conducted growing policy interventions. However, due to inefficient policy implementation, lack of scientific basis for policy making, and rapidly emerging environmental problems, the gap between the policies and their implementation is gradually widening.

Fifthly, the intertwining emergence of traditional environmental pollution problems and new environmental problems makes it more difficult to resolve these problems.

Except the above negative influence, there are a series of development trends that are favorable for the eco-environment.

Firstly, scientific and technological innovation and sharing economy will become important pillars for green development. Technological innovation in areas of Internet+, e-commerce platform, and big data facilitates the development of sharing economy. The sharing economy model for daily necessities and infrastructure related to the transportation, housing and other aspects will greatly enhance the resource and energy efficiency, reduce environmental load, and help form a more environmentally friendly lifestyle.

Secondly, integration of regional economy will further boost trade facilitation and help shape green industrial chain and value chain. Regional integration will enable more convenient exchanges of people and logistics. To promote the industrial system of green supply chain and shape green value chain will increase energy efficiency and reduce the negative impact on environment.

Thirdly, the green "Belt and Road" initiative and the financial agencies, for instance, Asian Infrastructure Investment Bank, which are promoted by China, will greatly enhance the environmental infrastructure construction and break the bottlenecks of environmental infrastructure in the region. The linkage of the "Belt and Road" initiative and the ASEAN Community construction will provide impetus and financial support for the regional green development. In addition, the green financial model will guide the capital investment to be allocated to the green industry and the new policies and means, such as green credit and green bond, will help prevent the capital-environment risks.

Fourthly, the implementation of the 2030 Agenda provides the strategic direction and cooperation impetus for the regional green development and South–South environmental cooperation. The 2030 Agenda marks the innovation of the global environmental governance model, with more comprehensive and systematic environment-related targets covering air, water, oceans, biodiversity conservation, and other aspects. The 2030 Agenda has a full-fledged monitoring system, in which the binding and quantitative indicators are made clear.

National action plans and regional cooperation are of great significance in promoting the realization of the 2030 Agenda. Thus, more efforts should be made to further explore modes for the South–South Environmental Cooperation and South–North–South Environmental Cooperation within the framework of the 2030 Agenda. China- ASEAN Environmental Cooperation has laid a good policy and practice foundation as a result of practice over the past years and built an important platform for future green development cooperation.

Apart from the above forecasted impacts and trends, the green development and environmental protection in China and ASEAN have their respective features.

(1) **China: A window of opportunity is presented in front of China for the comprehensive improvement of its eco-environmental quality. The future is promising for green development despite that there is still a long way to go**

- China still faces severe air, water and soil pollution. Emissions from production and daily life, urban and rural areas, industrial and transportation sectors would interweave with each other, and environmental problems become more complicated due to mixed pollution problems.
- Environmental protection's priority status on Chinese government's agenda will benefit multiple policy measures. During the 13th FYP period, pollution emissions are expected to reach the peak and decline by 2020,[10] resource and energy consumption would stay at a high level, yet their increase speed will slow down.
- The new-type urbanization which emphasizes optimized layout and resource/energy efficiency will create synergies with green and low-carbon economic growth. The "Belt and Road" initiative, Beijing–Tianjin–Hebei Economic Integration Strategy and Yangtze River Economic Belt Plan will provide golden opportunities for regional and international cooperation on environment and development.

The goal of the generally improved eco-environment quality is clarified in the 13th FYP. To achieve this goal, efforts should be made to firmly establish the development concepts of "innovation, coordination, green, open, and sharing," take the environment quality improvement as the core and ecological civilization system reform as the impetus, implement the most stringent environmental system, complete major tasks in the atmosphere, water, and soil pollution prevention, strictly prevent and control the environmental risks, strengthen the joint coordination in pollution control and ecological protection, accelerate the modernization of eco-environment governance system and capability, promote the multigoverning model, and adhere to the rule of law and environmental protection.

(2) **ASEAN: Positive policy signals and pragmatic actions will help solve environmental problems in the future**

- In light of fast economic globalization, the region will experience growing weight of overexploitation of natural resources and environmental pollution.
- Such environmental problems as forest coverage loss, water and air pollution, biodiversity loss will remain. If there is no adequate prompt and decisive measures, some places in the region would probably suffer from

[10]http://news.xinhuanet.com/energy/2015-01/29/c_127435090.htm.

water shortage by 2030.[11] Air pollution will cause higher health costs and production costs in many sectors.
- Lack of long-term effective management of environmental problems would likely threaten ASEAN's economic growth.
- Economic growth and improved environmental awareness enable environmental governance to release its potential. Regional environmental initiatives and mechanisms will help resolve regional environmental problems.

**Conclusions and Policy Recommendations: Establishing a Community for ASEAN-China Environmental Cooperation and Heading toward Green Development**

China and ASEAN Member States face severe resource and environmental challenges, and there is still a long way to go in green development. This not only requires all countries to further improve the policy framework, continue to innovate the policies and financing models at both the macro- and microlevels, and put their existing development, energy and environmental policies into real actions, but also requires the reinforcement of the international and regional cooperation by seizing the historic opportunity brought by the 2030 Agenda, linking the "Belt and Road" initiative with the ASEAN Community blueprint construction, continuing to strengthen the two parties' regional cooperation in green development, and expediting the green, low-carbon, and regional sustainable development.

China-ASEAN environmental cooperation needs to be further enriched under the newly approved five-year strategic cooperation framework, by detailing the action plans and developing the cooperation roadmaps. These concrete actions and green development practices will push forward the strategic green transition.

**Strategic Framework for China-ASEAN Environmental Cooperation**[12]

Based on the successful implementation of China-ASEAN Strategy on Environmental Protection Cooperation (2009–2015), both parties have jointly formulated the new ASEAN-China Strategy on Environmental Protection Cooperation (2016–2020) with the overall objective to strengthen China-ASEAN cooperation in priority areas of environmental protection by taking a coordinated and integrated approach with a view to achieve environmental sustainability in the region.

ASEAN-China Strategy on Environmental Protection Cooperation (2016–2020) specifies the following concrete objectives:

- Enhancing high-level policy dialogue with focus on environmental issues of common concern to increase understanding, enhance cooperation, and secure the harmonization of interests of ASEAN and China;
- Enhancing dialogue and cooperation on environmental protection;
- Improving capacity for national and regional environmental management through sharing knowledge and experiences and implementing joint actions;

[11]Asian Development Bank: ASEAN 2030: Marching toward A Borderless Economic Community.

[12]ASEAN-China Strategy on Environmental Protection Cooperation (2016–2020)

- Enhancing cooperation on priority areas, improving effectiveness and quality of cooperation, and developing good practices for regional and South–South[13] environmental cooperation;
- Supporting ASEAN Community's Post-2015 Vision.

**Priority Areas for China-ASEAN Cooperation**

(1) **Policy dialogue and exchanges**
Provide various platforms for China and ASEAN environmental decision makers to exchange ideas on key regional environmental issues and share environment management experience, take joint actions to improve the environmental cooperation, and implement the consensus reached by leaders in China and ASEAN.

(2) **Environmental data and information management**
Improve China's and ASEAN's capability of collecting, processing, and applying the environmental data and information. According to the cooperation initiative proposed at the 17th China-ASEAN Summit, the China-ASEAN environmental information sharing platform shall be established; knowledge and experience on environmental information and data shall be shared; and capacity building activities related to environmental information and data collection, processing, and application shall be carried out.

(3) **Environmental impact assessment (EIA)**
Improve China's and ASEAN's capability of environmental impact assessment by sharing the relevant knowledge and experience; conduct the capability building cooperation and the joint research on environmental impact assessment and management.

(4) **Biodiversity and ecological conservation**
Through the cooperation with the ASEAN Centre for Biodiversity, further develop and implement China-ASEAN Cooperation Plan on Biodiversity and Ecological Conservation, improve China's and ASEAN's capability and awareness in developing Biological Diversity conservation policies, strategies, or action plans, perform Convention on Biodiversity and other international obligations, and expedite the protection, management, and sustainable utilization of biological resources.

(5) **Promoting environmental industry and technology for green development**
Implement the ASEAN-China Cooperation Framework on Environmentally Sound Technology and Industry and support 10-Year Framework of Programmes on Sustainable Consumption and Production through establishing information exchange platform, carrying out demonstration projects and conducting joint research on environmental technology development.

[13]South–South Cooperation is a term generally used to describe the exchange of resources, technology and knowledge between developing countries.

(6) **Environmentally sustainable cities**
Improve China's and ASEAN's capability of promoting environmentally sustainable cities including small and growing urban areas, through knowledge and experience sharing as well as network and partnership building.

(7) **Environmental education and public awareness**
Support the implementation of ASEAN Environmental Education Action Plan (2014–2018) and enhance the public's environmental awareness through exchanges and cooperation among the environmental education institutions, relevant government departments, and civil social organizations in China and ASEAN Member States.

(8) **Institutional and human capacity building**
Improve the environment management capacity of China and ASEAN Member States by carrying out relevant capacity building activities under the Green Envoys Program. Strengthen the comprehensive training of environment management personnel; enhance policy formulation capacity in the areas of environmental economics and environment and health; establish experience sharing platform for environmental protection laws and regulation; and conduct mutual visits and exchanges for environment management staff to improve their capability.

(9) **Joint research**
Promote the communication between scholars and the think tanks and facilitate their capacity building so as to create green think tanks and support decision-making in China and ASEAN Member States. Study the global and regional emerging environment and development issues of common concern for both China and ASEAN and share the research achievements through existing China-ASEAN cooperation mechanisms so as to provide policy makers with the targeted, scientific, and information-based policy suggestions.

**Priority Actions and Practices for China-ASEAN Cooperation on Green Development**

Specific action plans should be designed for actions in the 9 priority areas under the framework set out by the ASEAN-China Environmental Cooperation Strategy. It will facilitate the translation of cooperation strategy into concrete actions and make unique contributions to the regional green development.

To carry out dialogues and exchanges on environmental policies. It is necessary to materialize the consensus reached by leaders of China and ASEAN via the existing high-level forum and ministerial conference. Actions include: update existing ASEAN-China Environmental Cooperation Forum to ASEAN-China Environment and Development Cooperation Forum; actively build the partnership between government, business, and the public, and establish a special cooperation mechanism; enhance the coordination and cooperation between the government, business, think tank and international organizations on environment and development in China and ASEAN Member States.

To strengthen environmental data and information sharing. Establish China-ASEAN environment and development information and service platform and database, incorporate the joint research achievements and China-ASEAN environment and development information into the database, and provide the information platform services through the Web site and other forms.

To improve biodiversity and ecological conservation. Improve China's and ASEAN's capability and awareness of developing biodiversity conservation policies, strategies, or action plans, implement Convention on Biological Diversity and other international obligations, and expedite the protection, management, and sustainable utilization of the biological resources. Highlight the sharing of urban and rural ecological protection experience and cooperate on the demonstration programs; explore the cooperation on biodiversity and ecological protection monitoring and demonstration programs; and carry out research on policy tools and practice related to biodiversity and ecological conservation.

To promote environmental industries and technologies for green development. Actions include: (1) to implement China-ASEAN Cooperation Framework on Environmentally Sound Technologies and Industries and support 10-Year Framework of Programmes on Sustainable Consumption and Production; (2) to establish China-ASEAN Cooperation Network on Environmental Technologies and Industries; (3) to establish China-ASEAN Platform of Information Exchange in Environmental Technologies and Industries and forge a platform of R&D and training; (4) to establish a batch of China-ASEAN demonstrative bases for international cooperation and exchanges of environmentally sound technologies and industries based on the existing bases in Yixing, Jiangshu Province and Wuzhou in Guangxi Zhuang Autonomous Region; and (5) to establish transfer markets to facilitate the development and transfer of environmentally sound technologies.

To construct environmentally friendly cities. Improve China's and ASEAN's capability of promoting environmentally sustainable cities including small and growing urban areas. Actively share the knowledge and experience on urban ecological protection, build the cooperation network, and promote the urban ecologically friendly development; enhance the cooperation on sustainable production and consumption in the context of urbanization; and cooperate on the environmentally sound treatment and disposal of the municipal waste.

To advance personnel exchange and capacity building. Improve the environment management capacity of China and ASEAN Member States by carrying out the relevant capacity building activities under the Green Envoys Program. Improve the capacity of environmental governance and green development of China and ASEAN Member States through knowledge sharing, training activities, and personnel exchange; carry out the "Green Envoys Program for the Maritime Silk Road" based on existing China-ASEAN Green Envoys Program, facilitating cooperation in personnel exchange, joint training and policy dialogue to improve capacities in the field of green development; and enhance capacity building in green development education and training.

To conduct the joint research on green development and green economy. Promote the communication between scholars and think tanks and facilitate their capacity building so as to create green think tanks in China and ASEAN. Conduct the joint research on green transition of industry, priority sequence in industrial transformation, green investment and trade, "21st Century Maritime Silk Road" infrastructure construction in a green way and conservation of environment and resources, resource efficiency improvement, recycling and remanufacturing, green finance, etc.; and conduct piloting and demonstration based on country specific situations.

To carry out demonstration programs to prepare for the best practice dissemination. Carry out demonstration programs on environmental technologies as well as environmental goods and services. Specific cooperation areas include: screening environment- friendly and clean production technologies; demonstration programs on pollution treatment technologies; cooperation and mutual recognition of environmental labeling (Eco-labeling) products; development of green industrial chain and supply chain; and other priority demonstration projects.

To promote sustainable consumption and production. Further facilitate the policy dialogues; carry out the joint research on such issues as national sustainable consumption and production strategy and roadmap, green supply chain and green procurement, etc.; and carry out pilots on environmental labeling and clean production.

Looking into the future, China and ASEAN Member States are expected to grow into key drivers for the economic development in Asia-Pacific region. It is up to current policies and decisions, however, to determine the development path and quality.

Existing experience in development around the world has clearly indicated the long-term benefits brought by green development. It is believed that through unremitting efforts, China and ASEAN Member States will be able to find an effective way to realize green development through policy innovation and practical actions, while realizing their economic development. Meanwhile, an attractive vision of green development of China and ASEAN is likely to lead the trend of regional green development and contribute to facilitating regional development.

# Chapter 1
# Overview of Green Development

**Abstract** This report defines green development as a new development model that is conducive to resource conservation and environment protection and established under the constraint conditions of resources and environmental carrying capacity. This chapter introduces the origination, definition, and features of green development as well as the practice and stories of typical countries and international organizations in their green transition. It also gives a systematic overview to the international and China-ASEAN evaluating indicator system for green development and introduces the overall structure of this report.

## 1.1 Background of Green Development

In the traditional development model, the huge consumption of mineral energy subjects the physical environment of human society to severe environmental pollution and resource exhaustion which results in an overall ecological crisis. These increasingly threaten the survival and future development of human, thus stimulate constant reflections on the prevailing economic development model.

The shortcomings of the traditional economic development model were not quite evident until the 1960s. In 1961, the United Nations approved the United Nations Development Decade, stating that economic growth does not necessarily represent development, because economic growth demonstrates an increase in output with the use of resources while development is defined by improvements in quality. Approved in 1972, the Declaration on the Human Environment ushered in a new era of environmental protection. In the 1980s, World Commission on Environment and Development (WCED) released "Our Common Future," defining sustainable development as "development that meets the needs of the present without compromising the ability of future generations to meet their own needs." In the 1990s, the United Nations Conference on Environment and Development (UNCED) approved the Rio Declaration and Agenda 21, which made clear the "common but differentiated responsibilities" that developed countries and developing countries should perform in coping with global environmental issues. In December 1997, the

 
China-ASEAN Environmental Cooperation Center, *China-ASEAN Environment Outlook 1 (CAEO-1)*, https://doi.org/10.1007/978-981-10-6211-7_1

3rd Conference of the Parties to the United Nations Framework Convention on Climate Change (UNFCCC) adopted the Kyoto Protocol, specifying the emission reduction targets of developing countries, with the aim of restricting their emissions of greenhouse gases. From the presentation of the sustainable development concept to the formulation of a group of worldwide regulations for environmental protection, it proves human beings have become evidently aware of the disadvantages of the traditional development model and have started to develop more resource efficient and environmentally friendly production and consumption alternatives. In the twenty-first century, world development highlights human development which treats green development as the major theme. The concept of green development, derived from sustainable development, becomes increasingly a focus of discussions of international organizations. In 2002, the United Nations Development Programme (UNDP) issued a report titled China Human Development Report 2002: Making Green Development A Choice, preferring green development.

To alleviate the impact of the financial crisis in 2008, promote the rapid recovery of economy, and bring forth new job opportunities, countries and organizations worldwide have taken measures to facilitate green economic development, formulate green growth strategies, build a green society, strengthen green investment and initiate green action plans. The Global Green New Deal put forward by the United Nations Environment Programme (UNEP) is one of such green development strategies.

**Box 1.1 Five Priority Sectors Underpinning a Global Green New Deal**

- Clean energy and clean technologies, including recycling;
- Rural energy, including renewables energy and sustainable biomass;
- Sustainable agriculture, including organic agriculture;
- Ecosystem infrastructure;
- Reduced emissions from Deforestation and Forest Degradation (REDD).

Source: United Nations Environment Programme (UNEP), 2008

In April 2009, state heads worldwide signed an important agenda for the "inclusive, green and sustainable economic recovery" on the G20 London Summit. Moreover, the Organization for Economic Cooperation and Development (OECD) issued a report on green growth strategy in 2010. In 2012, the "Rio + 20" Conference highlighted the theme of green development, with the aim to promote the realization of comprehensive, balanced, concerted, and sustainable development through overall planning and coordination of economic, social, and environmental factors [1].

## 1.2 Concept and Characteristics of Green Development

### *1.2.1 Concept of Green Development*

At present, green development may be defined from two perspectives: [2, 3]

(1) Emphasize that in the course of economic development, attention should be paid to the reduction in greenhouse gas emissions and the strengthening of resource and environmental protection, with the logical aim to address issues and challenges related to climate change and conserve resources and the environment;
(2) Emphasize that green emerging industries should be taken as a new economic growth point and a driver for clean economic growth, with the ultimate purpose to boost economic growth. For instance, the South Korean Government (2010) regards "green growth" as a new development model where new growth impetus and job opportunities are brought forth by green technology and green energy.

Countries have different interpretations about green development owing to the differences in terms of development phase and geographic location. Developed countries in the post-industrialization phase have largely removed traditional environmental pollution problems, so they focus more on such global environmental issues as climate change as well as the establishment of institutional frameworks that aim to solve these problems and the cooperative actions among countries. However, China lays particular emphasis on the development of domestic economy and ecological and environmental protection, rather than social inclusiveness, for it stays under industrialization and has yet solved traditional environmental pollution problems.

Green development is an evolving concept and continues to grow as economy and society develop and environment and resource restrictions and pressure increase. The report offers the following definition:

Green development is a model innovation based on the traditional development, the constant doubt, and reflection by human being about traditional models of industrialization and urbanization, and a new development model, contrary to the traditional "black" development model and in favor of resource conservation and environmental protection, established under the constraint conditions of environmental capacity and resources carrying capacity (Table 1.1).

With the overall aim to realize the coordination between economic development and resources and the environment, green development works to reduce the impact on resources and the environment while seeking economic growth. There are three objectives:

- To address preferentially the problems of resources and environment facing the mankind through energy conservation and emission reduction;
- To improve the resource efficiency and green competitiveness of industries by virtue of technical progress; and
- To transform economic development mode and promote the green development of the economic system by means of green transition.

**Table 1.1** Differentiation between green development, sustainable development, green economy, circular economy, and low-carbon economy

| Item | Sustainable development | Green economy | Circular economy | Low-carbon economy |
|---|---|---|---|---|
| Content | Sustainable development meets the demands of the contemporary generation without compromising those of the future. It has richer implications than green development, and green development is one of the most effective approaches toward sustainable development | Green economy is a kind of economic pattern that highlights greening economic activities from industrial economy point of view. Its three key areas are green investment, green construction, and green growth | Circular economy includes reduction, reuse, and recycle. Green development is a wider concept than circular economy. Circular economy can be regarded both as a means toward and an integral part of green economy | Low-carbon economy was proposed to address global energy security and climate change. The focus lies in reducing carbon emission intensity and controlling the growth of $CO_2$ emission |
| Similarities | Green development and sustainable development both prioritize scientific development, with fundamental objectives to improve people's living environment and standard. They emphasize that the economic and social development should be compatible with the carrying capacity of resources and environment | The objective of green development and green economy is both to balance economy and environment and require a change of economic development pattern, adjust industrial structure and reduce the discharge of pollutants. Both need to harness the market mechanism, advance scientific and technological innovation, and expand public participation | Green development and circular economy both stress resource conservation, reduce the discharge of pollutants, and recycle wastes | Green development and low-carbon development both stress the change of the development mind-set of traditional industrial society and the integration of sustainable development into the whole economic activities of human being |

(continued)

**Table 1.1** (continued)

| Item | Sustainable development | Green economy | Circular economy | Low-carbon economy |
|---|---|---|---|---|
| Differences | Sustainable development is passive in adapting to the natural constraints, but green development is proactive. Sustainable development is about not to leave regret for future generation, while green development is more about green input to leave more ecological capital | Green development incorporates more considerations of social progress such as social justice, education equality, and poverty. It has wider implications than green economy | Circular economy emphasizes more on resource recycling and reuse | Low-carbon economy is largely targeted at energy and carbon emission |

### *1.2.2 Characteristics of Green Development*

In essence, green development features energy-saving and environment-friendly development [4]. It is directed at minimizing environmental load through available and appropriate technologies and economic conditions, and finally controls such load and impact within the resource supply capacity and environmental self-purification capacity of the ecosystem [5] and brings forth a virtuous cycle between social and economic activities and ecosystem. Specifically, the characteristics of green development include: [6]

- Low consumption;
- Low emission;
- Low pollution;
- High benefit;
- High-carbon sink;
- High cycle.

It should be noted that green development stresses that there is no "universally applicable" green economic transition model for all countries. Green development should be translated gradually into a differentiated or regionalized growth model, taking into consideration the limiting conditions and opportunities of each region.

## 1.2.3 *Approaches Toward Green Development*

### 1.2.3.1 Transform Economic Development Mode and Adjust Industrial Structure [7]

A stable environment and a constituency that commits to maintain a balanced ecology are the foundations of any form of development. The transformation from traditional and unsustainable development toward green development requires improved capacities to understand and apply principles of ecology, sustainable development and enhance social responsibility. A green economy contributes to a sustainable global economy and, in the process, increases green job opportunities, encourages innovations in business model, and drives the development of green products and services.

### 1.2.3.2 Promote Sustainable Consumption of Resources and Energy and Intensify Pollution Control

The key of green development is to optimize resources and energy mix and improve their utilization efficiency. Such facts as high proportion of fossil fuels, low energy use efficiency, high energy consumption intensity, low integrated use efficiency, high resource consumption, and high pollution are bottlenecks restricting the green development of national economy. Therefore, to decrease the consumption of resources and energy, improve their utilization efficiency, and intensify pollution control is among the important approaches for realizing green development.

### 1.2.3.3 Accelerate the Independent Innovation and Application of Green Technology Using Market Mechanisms

Green technology supports green development. Many developed countries have strived to develop green technology in such fields as new energy, with the aim to take the lead in the new round of technological and industrial reform and thereby ensure their respective technological competitiveness stays at a leading position in the world as well as the top of world economic division of labor and global product chain.

### 1.2.3.4 Establish Economic Policies, Laws and Regulations in Favor of Green Development

To evaluate and internalize the values of ecosystems, including those of resources and environment, is the fundamental way to remove the disadvantages of traditional development, promote green transition, and realize green development. To implement green development strategy, governments must transform the economic cost of the society incurred by producers back into their private cost through market, legal, and policy means and encourage producers to make innovation in green technology and perform green transition. Moreover, governments should accelerate the ecological innovation of traditional industrial standards, develop and innovate in pertinent standards consistent with sustainable development, and establish a green standard system covering all fields.

#### 1.2.3.5 Advocate the Concept of Green Development and Raise Public Awareness

The limited uptake of a low-carbon lifestyle is due to the poor understanding and backward sense of environmental protection and green development. There is thus a need to improve efforts in environmental education and publicize the significance and necessity of a low-carbon lifestyle among several stakeholder level. Efforts should be made to guide consumers through proper policy measures to practice low-carbon life and help children understand environmental protection in each respect of life.

## 1.3 International Progress in Green Development

### *1.3.1 Green Transition Courses of Major Countries and International Organizations*

Green development has been promoted to be a national strategy in many countries. Various countries are emphasizing to incorporate ecological concepts in economic development and adopt approaches toward green development. In recent years, such economies as USA, EU countries, and Japan have put forward the green development strategy, implemented "green new deal," and developed green economy rapidly as a response to issues related to economic recovery and climate change.

#### 1.3.1.1 UK, Germany, and France Lead in Green Development

In UK, the development of green energy is taken as the priority of green economy policy. On July 15, 2009, UK issued Low Carbon Transition Plan and Renewable Energy Strategy, which can be regarded as the most systematic government white papers of developed countries for coping with climate change. It sets UK up as the first country that identifies carbon emission management plan in government budget framework. In August 2011, UK published "Green Economic Transition Plan:

Government-Enterprise Cooperation," stating that 45–60% of UK petroleum consumption and 70% or more of its natural gas consumption will be imported by the year 2020.

In Germany, the emphasis of green economy lies in the development of ecological industries. In June 2009, Germany published a strategic document aiming to promote its economic modernization. In this document, the German government stressed the ecological industry policy should serve as a guideline for the country's economy. Following that, the government issued an 11 billion US dollars economic stimulus package in 2009 and 2010 to continue implementing energy-saving reconstruction of buildings. After the Fukushima nuclear disaster, Germany determined to stop nuclear energy development, ushering in an era of new energy oriented to wind power and solar power. As for the Energy Act, the long-term strategic purpose is to make Germany a world leader with regard to energy efficiency and green economy.

In France, green economy policy focuses on the development of nuclear energy and renewable energy. In December 2008, the Ministry of Environment announced a package plan aiming to develop renewable energy. In addition to the great efforts to develop renewable energy, the French government spent 400 million Euros in 2009 developing clean energy vehicles and "low-carbon vehicles." Moreover, nuclear energy has been always the focus of France's energy policies as well as the emphasis of its green economy. According to France's Energy Transition for Green Growth act, the government plans to reduce the share of nuclear power in the national energy mix from 75 to 50% by 2025 and raise the share of renewable energies in final energy consumption from 13.7 to 32% by 2030.

#### 1.3.1.2 USA, Japan, and South Korea Promote the Green New Deal Vigorously

In the USA, the "Green New Deal" involves such fields as energy conservation and efficiency improvement, new energy development and coping with climate change. New energy development plays the core part. Up to February 15, 2009, the American Recovery and Reinvestment Act, with a total investment of 787 billion US dollars, took new energy development as a main field; meanwhile, the USA exerted great efforts to propel the development of fuel-efficient cars and green buildings and proactively encouraged the development of emerging industries. In December 2012, President Obama declared the "Better Building Initiative," which has "Better Buildings Challenge" as a part of it, with the aim to reduce the country's energy consumption for building by nearly 40 billion US dollars in the next decade.

In April 2009, the Japanese government published a policy draft titled Green Economic and Social Transformation, for the purpose of enhancing Japan's green economy by such measures as reducing greenhouse gas emissions. In May 2009, Japan initiated the Eco-Point system that supports energy-saving household appliances, making energy conservation a main social sense through daily consumption and demonstrating the social influence of green economy in a concentrated manner. In the meantime, Japan took the lead in presenting the idea of building a low-carbon society, declared to guide the world's low-carbon economy reform and proposed making Japan the first green low-carbon society. In April 2014, Japan formulated the new Basic Energy Plan, stating clearly that efforts should be made to minimize the nuclear power's proportion in the energy mix by means of energy conservation, renewable energy development, and improvement of thermal power efficiency. In July 2015, Japan's Commitment (Draft) confirmed the objective of reducing greenhouse gas emission by 26% from 2013 to 2030.

South Korea plans to reproduce the Miracle on the Han River by virtue of the green growth strategy. In early January 2009, the South Korean government put forward a strategy of promoting economic revival through low-carbon green growth, which advocated creating job opportunities and stimulating economic growth through investing in green ecological projects. In July, South Korea announced the national green growth strategy and five-year plan, confirming a group of indexes of "green energy." Solar energy industry and smart power grid are advantageous green environment-friendly industries that have emerged in recent years. Since the beginning of 2011, South Korea has kept improving the electric car-charging equipment in cities so as to meet the need of electric car development. The South Korean government hopes its popularizing rate of electric cars can reach 20% by the year 2020.

OECD member countries also approved the ministerial council declaration in June 2009 to advocate "green growth," i.e., promoting economic revival via policy tools and green investment in a short term and propelling sustainable growth through building environment-friendly infrastructure for green economy in a long term while transiting to a sustainable low-carbon economy. In March 2009, EU announced it would spend 105 billion Euros prior to 2013 supporting the promotion of "green economy plan" in various countries, of which 54 billion Euros would be used to help the countries to implement EU environmental protection laws and regulations and 28 billion Euros to improve waste treatment technology and water quality.

In recent years, developed countries have proactively resorted to green development to promote economic growth, and emerging market economies have obtained remarkable achievements in popularizing green development using their respective advantages. Green economy has shown new characteristics in the global development, such policies and regulations as carbon tax, carbon labeling, and green procurement have grown increasingly influential, and new energy industry development has slowed down after the rapid growth. Countries are more willing to cooperate with one another in green development, despite of the outstanding differences, the competition over the dominant right grows increasingly fierce, and green protectionism improves further. As the European debt crisis continues, developed countries witness a decline in economic growth, and emerging economies are generally confronted with demand contraction, insufficient growth impetus, green development has served as an important guide for the development transition and upgrading of China and other emerging economies.

### *1.3.2 Major Characteristics, Experiences, and Inspirations of International Green Development*

The emphasis on promoting green development lies in structural reform, including the greening economic structure, industrial structure, and energy structure. Efforts should be made to better promote green development as well as the concerted development of economy and environment through structural reform. International experience in this respect shows the following important facts and common characteristics:

- Tertiary industry (service-oriented) sectors account for an increasingly bigger share in the economy. Moreover, the "service-oriented extension" of industrial sectors starts gradually to play an important role, as the additional value is created in production and within product life cycle;
- The diversification and professionalization demonstrated in production flows and centering on production links have served as a driving force for structural adjustment and upgrading, which gives rise to the global labor force polarization and transfer in production activities;
- It has been a part of structural reform to transfer outdated production capacity and industries from developed industrialized economies to underdeveloped or developing countries, through export-oriented foreign direct investment.
- The following experience of various countries, represented by major developed countries, in green transition of industries is worth learning:
- Both government guidance and market operation should be emphasized in the green transition of industries;
- The formulation and implementation of industrial transformation policies is of great importance, and to promote the green transition of industrial structure should be taken as a fundamental measure;
- Technological research, development, and innovation are an important driver for the green transition of industries.

## 1.4 Green Development Assessment Indicators

In recent years, achievements have been obtained constantly in the research of green development assessment index system. These research efforts focus on three approaches, including green national economic accounting, multi-indicator measurement system of green development, and green development assessment indicator system [8–11].

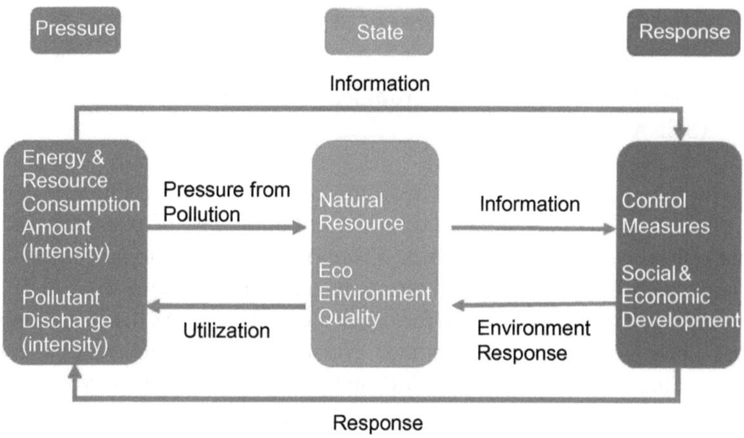


**Fig. 1.1** PSR Model

### *1.4.1 Green Development Assessment Models*

In terms of green development assessment, the "pressure-state-response" or PSR model is a typical one, which was developed by OECD and UNEP at the end of 1980s to assess resource utilization and sustainable development. P represents human being's economic and production activities and consumption model which incurs unsustainability. S represents the system state in the sustainable development process. R represents the response measures that human being takes to promote sustainable development. This model can reflect the interactive relationship between nature, economy, and society in a good way and is widely adopted in different areas as it is highly systematic and applicable to different spatial dimension and domain with rational categorization of different indicators based on the interaction between economy, society, and environment. Environmental quality and natural resource state is the State of PSR model. Social and economic development and the measures taken by human society in response to the problems are the Response of the model and the pressure brought by human activities to environment is the Pressure. They constitute the PSR model for green development assessment (see Fig. 1.1).

Green development assessment indicators can be established by referring to the three modules of PSR based on its cause–effect relationship. Users can localize indicators that are specific to their own conditions under the PSR framework to gauge their progress in green development. Different countries can adopt country-specific indicators. This report establishes a China-specific indicator system as an example (see Table 1.2).

## 1.4.2 *Global Green Development Assessment Systems*

### 1.4.2.1 System of Integrated Environment and Economic Accounting (SEEA)

The traditional national economy accounts can reflect the economic aggregate alone, rather than the consumption cost and pollution cost of resources and environment caused by economic activities. Therefore, the international community has started to establish a green national economy accounting system so as to offset the defect of the distorted economic performance measurement. As the first country that practices natural resource accounting, Norway released its natural resource accounting data for the first time in 1981 and published a report on Norway natural resource accounting in 1987. By reference to the experience of Norway, Finland established the natural resource accounting frame system covering forest resources, environmental protection expenditure, and air pollutant emission. In 1993, United Nations Statistical Office incorporated resources and environment statistics into the national accounting system, presenting a system framework for interpreting environment and resource reserves and flows consistent with the System of National Accounts (SNA), namely the System of Environment-Economic Accounting (SEEA), which provides a theoretical framework for various countries to establish a green national economy accounting system. Based on the practice of various countries, UN has made a constant improvement and released SEEA 2000 and SEEA 2003 successively, providing reliable guidance and security for further standardizing the green national economy accounting system of the countries.

Once released, SEEA has attracted great attention of the international community and it's among the green national economy accounting systems that are most influential and applied most widely. Subsequently, there has been national economic accounting matrix with environment accounts (NAMEA), European system for the collection of economic information on the environment (SERIEE), environment and natural resources accounting project (EARAP), etc. [12].

### 1.4.2.2 OECD Green Growth Indicators [13]

In recent years, OECD states have recognized gradually that development and environment are mutually supportive and believed efforts should be made to implement the green growth strategy so as to cope with the financial crisis and give a policy response for a longer term. Hence, at Meeting of the OECD Council at Ministerial Level held in 2009, the ministers from 34 countries inspired OECD to formulate a green growth strategy and draw up a comprehensive framework integrating economy, environment, society, technology, and development for the countries.

According to OECD understanding, green growth aims to achieve economic growth. It stresses to address such problems as environmental degradation and

**Table 1.2** PSR indicator system for green development assessment (with China as an example)

| Item | Element | Indicators |
|---|---|---|
| Pressure | Pollutant discharge | Emissions of major pollutants per unit GDP ($SO_2$, $NO_x$, VOC, COD, ammonia nitrogen) |
| | | Fertilizer use intensity |
| | Resource consumption | Water consumption per unit of GDP |
| | | Energy use per unit of GDP |
| | | Land development and utilization intensity |
| State | Environmental quality | Average annual concentration of major pollutants ($PM_{2.5}$, $SO_2$, $NO_x$, $O_3$, COD, ammonia nitrogen) |
| | | Percentage of the days with air quality above Grade-II |
| | | Percentage of provincial monitoring section with water quality above Grade-III standard |
| | | Coverage ratio of noise standard-reaching areas |
| | Ecological state | Forest coverage |
| | | Percentage of wetland to national land area |
| Response | Economic development | Percentage of R&D input in GDP |
| | | Percentage of the output value of "high-energy consuming, high polluting, and resource-based" industries in the industries above scale |
| | | Per capita GDP |
| | Government management | Reuse rate of reclaimed water |
| | | Centralized treatment rate of urban life sewage |
| | | Non-hazard treatment rate of urban life wastes |
| | | Per capita park greenland area |
| | | Share of urban public transit in people's transportation |
| | | Control rate of village environment |
| | | Share of environment protection input in GDP |
| | Public response | Degree of satisfaction of environmental quality |

climate change from the perspective of economy, which is fully expressed in the green growth strategy framework it establishes (see Fig. 1.2). Based on the framework, OECD has built up a complete green growth indicator system that covers economy, environment, human well-being, etc., and contains four first-grade indicators, including environmental and resource productivity, natural asset base, environmental dimension of quality of life, and economic opportunities and policy responses, 14 second-grade indicators and 23 third-grade indicators.

#### 1.4.2.3 UNEP Green Economy Progress Measurement Index System [14]

The international community has achieved significant progress in the establishment of sustainable development indexes since the UNCED reached the consensus of

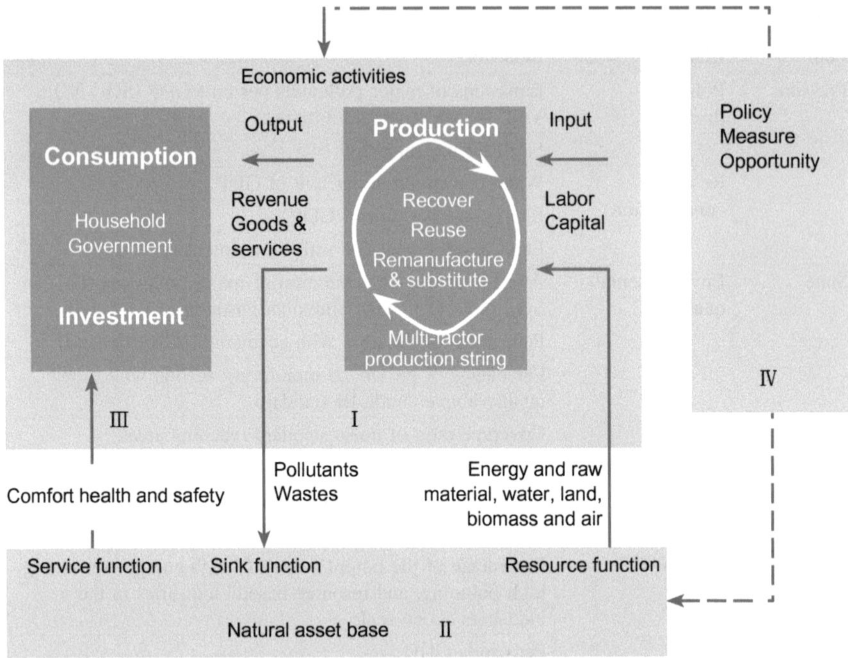


**Fig. 1.2** OECD green growth strategy framework

sustainable development in 1992. However, there are still serious challenges. In recent years, green economy has been a new trend in the sustainable development field. Therefore, UNEP starts to conduct researches to build a green economy progress measurement framework for providing timely and practical guidance for the policy makers of various countries. UNEP green economy progress measurement framework covers mainly economic transition, resource efficiency, and social progress and human well-being (see Fig. 1.3).

In addition to the above-mentioned OECD and UNEP systems, the multi-indicator measurement systems also include the following:

The UNESCAP ecological efficiency index system, established by the United Nations Economic and Social Commission for Asia and the Pacific (UNESCAP), is designed to reflect the interaction and relations between economic activities and resources and environment, with the purpose to assess how environmental development of the Asian-Pacific Region influences the sustainable development of economy and society.

The California green innovation measurement system is used to monitor the overall development of green economy in California, including transportation and recoverable energy, and measure how the economy of California develops through green innovation. The green innovation measurement system 2012 is composed of low-carbon economic system, energy efficiency system, green technological

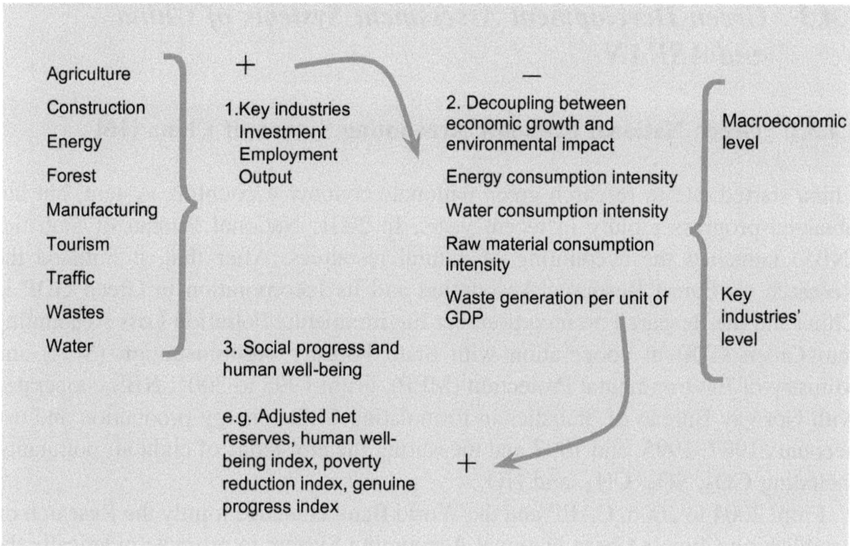


**Fig. 1.3** UNEP green economy measurement framework

innovation system, renewable energy system, and transportation system and contains 18 sub-indicators.

#### 1.4.2.4 Environmental Performance Index (EPI) Put Forward by Yale University, et al. Yale University and Columbia University Published Jointly the Environmental

Sustainability Index (ESI), to provide quantitative index for assessing sustainable development, and support the millennium development goals. To assess the countries' achievements in environmental improvement preferentially and provide them with policy guidance, Yale University and Columbia University Joint Study Group issued the Global EPI Report based on ESI on the World Economic Forum in 2006, 2008, and 2010, respectively, with important influence on the international community.

EPI centers on two basic goals of environmental protection reducing environmental pressure on human health and improving ecosystem vitality and promoting the good management of natural resources. These goals cover 22 indexes in ten fields, including climate change, agriculture, fishery, forest, water source, air pollution, and environmental burden.

## *1.4.3 Green Development Assessment Systems of China and ASEAN*

### 1.4.3.1 Green National Economy Accounting System of China [10]

China started late to research green national economy accounting system, but has obtained progress rapidly in recent years. In 2001, National Bureau of Statistics (NBS) launched the accounting of natural resources. After that, it initiated the Research on Forest Resource Accounting and Its Incorporation in Green GDP in China and the Research on Incorporating Environmental Pollution Loss Accounting into Green GDP in cooperation with State Forestry Administration (SFA) and Ministry of Environmental Protection (MEP). From 1998 to 2001, NBS cooperated with Norway Bureau of Statistics in formulating China energy production and use accounts 1987, 1995, and 1997 and measuring the emissions of eight air pollutants, including $CO_2$, $SO_2$, $CH_4$, and $NO_x$.

From 2004 to 2006, CAEP and the World Bank launched jointly the Research on Establishing China's Green National Accounting System to promote practically the application of the green national economy accounting system in China, and completed such important researches as China Green National Economy Accounting System Framework, and China Environment and Economic Accounting Research Report for seven consecutive years.

### 1.4.3.2 Resource and Environmental Performance Index Put Forward by CAS Sustainable Development Strategy Research Group

In 2006, CAS Sustainable Development Strategy Research Group proposed the resource and environmental performance index (REPI) and monitored and comprehensively evaluated the resource consumption and pollution emission performance of countries and regions. At state level, REPI chooses four resource consumption intensity indexes including energy consumption intensity and three pollutant emission intensity indexes including COD emission per unit of GDP, and comprehensively assesses the recourse and environmental performance of all provinces (municipalities and autonomous regions) from 2000 to 2009 through equal-weight assignment method, so as to reflect the relative level of resource utilization technology and the relative pressure of economic development on resources and environment of each province in China.

### 1.4.3.3 Green Development Index Report Made by Beijing Normal University [15]

Beijing Normal University (BNU) establishes the index system from the perspectives of the green degree of economic growth, the potential capacity of resources and environment, and the support of government policies and chooses nine second-grade indexes and 55 basic indexes. The index system can reflect roughly the green development level of China. China's green development index 2012 is composed of China inter-provincial green development index and China urban green development index, used, respectively, to measure and rank the green development of 30 provinces, municipalities and regions, and 38 large cities.

#### 1.4.3.4 Green National Economic Accounting System of the Philippines

The Philippines has two different green national economic accounting systems, namely Environmental & Natural Resources Accounting Project (ENRAP) and SEEA. The Philippines began to develop its accounting system ever since 1990. ENRAP was developed by Peskin and sponsored by USAID. It initially was to look at the depletion of forest resources and was expanded to other natural resources including fishery, mining industry, and soil. In addition, its National Statistical Coordination Board (NSCB) developed Philippine SEEA (PSEEA) in 1995. Though ENRAP and PSEEA have similarities, the two are still quite different as a result of respective theoretical base.

**PSEEA System**

PSEEA was largely built on SEEA. NSCB has developed three natural resource accounts for forest, fishery, and mining resources, respectively, and uses net price to estimate natural resource consumption. Capital accounts include resource initial stock, stock change due to economic activities (e.g., depletion or exploration), other aggregates and volume change due to other factors (e.g., technology progress or discovery of new mines) and closing stock. PSEEA then takes all these information to establish capital account with currency value.

PSEEA uses cost maintenance method to estimate environment degradation with the minimum cost of available equipment to calculate. The cost includes annualized capital cost, maintenance cost, and operating cost.

As PSEEA only calculates the services provided by the natural environmental resource with market trading price, those environment services without market price is not included.

**ENRAP System**

ENRAP was put forward by US economist Prof. Peskin. As traditional national economic accounting system only covers those resources with market trading price and thus ignores the environmental services without market price and environmental quality degradation caused by pollution, ENRAP proposes to incorporate all those into the green national economic accounting system.

ENRAP uses the method in line with economic theory to calculate the consumption of natural resources and environmental pollution and damage. It is established on the precondition that all the economic input and output in the economic system shall be recorded in the accounting tables.

ENRAP includes the services provided by the asset with market price, but for those non-monetized services, it uses shadow price for estimation. There are three major services, including athe input service by the environmental resources, e.g., waste management; bthe output service, e.g., leisure service; and closses caused by pollution. Though ENRAP tries to monetize the environment, Peskin also admits that not all the environmental assets can be successfully monetized. Therefore, ENRAP also has some physical accounts.

ENRAP also has three new items, namely waste treatment service by the environment, environment loss, and the direct service by the environment. ENRAP also defines the indicators of net environmental benefit or NEB which is the difference of environment service value (sum of waste treatment service and direct service) and environment pollution loss. It can be used as a rough indicator to estimate environment management efficiency. If NEB is 0, it means environment management has reached an ideal state. If NEB value is lower than 0, it means further improvements are needed.

ENRAP adopts “willingness to pay” to measure how satisfied the public is toward various environmental services. In terms of environment losses, it also takes the public’s “willingness to pay” to avoid causing environmental losses as the evidence. The Philippines also uses does–response relationship to gauge the impact of environmental pollution to human health and has it monetized. Results show that local water pollution is more serious than air pollution. On natural resources, ENRAP uses net present value method to calculate the consumed value of natural resources like forest, fish stock, minerals, and soil.

On air pollution, ENRAP covers $PM_{10}$, $SO_x$, $NO_x$, VOC, and CO. On water pollution, it takes $BOD_5$ as the parameter. Water pollution sources include household wastewater and rainwater runoff polluted by nitrogen in agricultural and forestry sectors.

#### 1.4.3.5 Other Green Development Assessment Systems

In 2008, China’s urban green development assessment system was built up by the Organizing Committee of China International Forum on Green Development. Composed of six objective indexes and four other indexes, the system covers public assessment of environmental quality, air quality, waste treatment, efficient use of water, efficient use of energy, urban environment improvement, urban afforestation, environmental improvement investment, public environmental demand, local government’s enforcement of the environmental laws, regulations, and policies issued by the central government, and the expert evaluation of overall city image. It is used to assess the green development of 656 cities in China, including 369 county-level cities, 268 prefecture-level cities, 15 vice provincial-level cities, and 4 cities directly under the central government.

To learn about the low-carbon economic development of each region in China, the Renmin University of China published a provincial low-carbon economic competitiveness index system consistent with China's national conditions in 2011. Taking three core factors, i.e., low-carbon efficiency, low-carbon guidance, and low-carbon society, as the first-grade indexes and containing eight second-grade indexes and 21 third-grade indexes, the system presents a comprehensive picture of the low-carbon development level of each region in China and is used to rank the low-carbon competitiveness of all provinces, municipalities, and autonomous regions between 2005 and 2009. On this basis, China is divided into low-carbon region, medium-carbon region, and high-carbon region, and these regions are compared with one another.

In 2012, Beijing Technology and Business University and partner research institutions established green economy indexes starting with resource and environmental efficiency, taking Beijing as the benchmark city of green economy index and the resource and environmental efficiency of Beijing as the base to calculate the green economic development level of other cities. In fact, it is a simplified version of ecological efficiency assessment index system and, thereby, is more operable in assessing the resource and environmental efficiency and green economy index of 273 cities and 31 provinces, municipalities, and autonomous regions in China.

In 2012, China Center for International Economic Exchanges (CCIEE) and World Wildlife Fund (WWF) proposed that green economy index system be conceived in three dimensions, including social and economic development, resource and environmental sustainability, and green transition driving, which are also first-grade indexes. There are six second-grade indexes, 14 third-grade indexes, and 30 fourth-grade indexes in their index system.

In 2014, Jiangsu Provincial Academy of Environmental Sciences (JSAES) and Chinese Academy for Environmental Planning (CAEP) completed jointly the green development assessment of Jiangsu Province for 2010–2012, establishing the green development assessment index system of Jiangsu Province and assessing the green development course of the province and the municipalities directly under the provincial government from the perspectives of resource utilization efficiency, environmental protection cost, and contribution of ecological products [2]. The system consists of 11 third-grade indexes. Weighting of the indexes considers the actual situation of Jiangsu Province, and the model is built through the target value-scoring method. To make the indexes comparable, ratio method, segmented assignment method, and linear interpolation method are combined to calculate the scores of each index. This is the first green development assessment report that the Chinese government issued according to the overall requirement of "incorporating resource consumption, environmental damage and ecological benefit into the economic and social development assessment system" as specified in the report of the 18th CPC National People's Congress. At present, the assessment has been conducted to the year 2014.

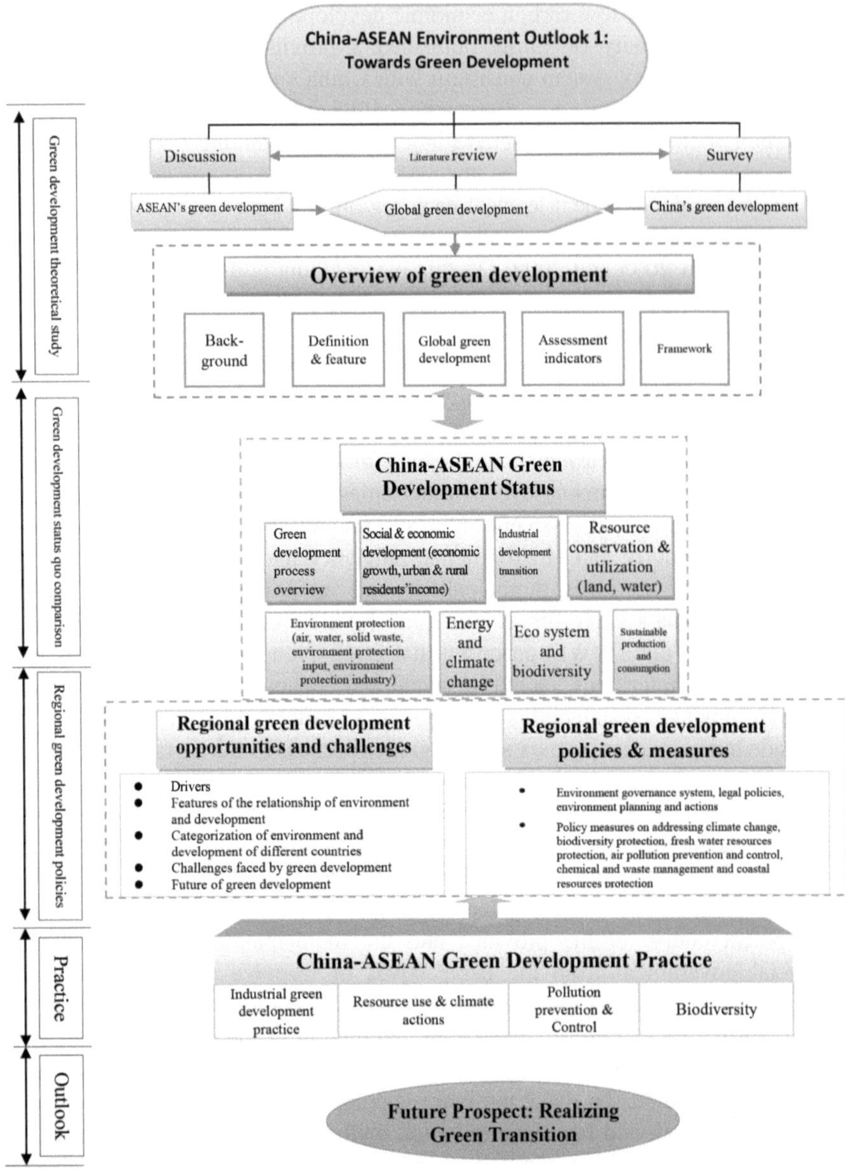


**Fig. 1.4** Overall structure and research framework for CAEO-1

## 1.5 Overall Structure and Research Framework for CAEO-1

This first edition of China-ASEAN Environment Outlook (CAEO-1) reviews the background of green development at the very beginning and then puts forward its definition of green development, followed by discussions of green development practices of different countries. It elaborates and compares the green development status of China and ASEAN. Based on analysis of environment–development relationship, CAEO-1 describes development features of typical countries in the region and the corresponding challenges and opportunities. It gives a snapshot review of the policy measures of China and ASEAN Member States (AMS) on green development and environment and in-depth analysis of typical policy measures, policy effects, and constraints. It probes into selected case studies on green industrial development, resource use and climate actions, pollution prevention and control, as well as biodiversity from China and AMSs such as Thailand, Singapore, Viet Nam, the Philippines, and Cambodia. In the end, CAEO-1 puts forward policy recommendations for promoting green development cooperation within the region (Fig. 1.4).

## References

1. Zhang, M.: Green development: international progress and China's pathway [J]. Int. Stud. **5**, 93–102 (2013)
2. Chinese Academy for Environmental Planning, Jiangsu Provincial Academy of Environmental Science, Nanjing University. Jiangsu Green Development Report [R], (2014)
3. Lan, Z.: Strategies of green development in China [J]. Ecol. Econ. **3**, 80–83 (2008)
4. Wang, J., Cao, D., Chen, X.: Conception of national green strategic plan [J]. Environ. Prot. **6**, 39–43 (2006)
5. Zhu, D.: New concept of green economy and consideration of deepening green economy studies in China [J]. China Popul. Resour. Environ. **22**(5), 40–47 (2012)
6. Liu, Q., Shen, H.: Study on index system and assessment method for regional sustainable development [J]. China Popul. Resour. Environ. **4**, 60–64 (1997)
7. Hongchun, Z.: Take the road of low-carbon green development with Chinese Characteristics [J]. Recycl. Resour. Circ. Econ. **04**(6), 4–9 (2011)
8. Qin C., Wang H., Tian Y., et al.: Study on evaluation indexes of resources and environmental carrying capability [J]. China Popul. Resour. Environ. **S2**, 335–338 (2011)
9. Cao, J.: Environmental and economic accounting [M]. Economy & Management Publishing House, Beijing (2004)
10. Wang, J., Cao, D., Jiang, H.: Green national accounting for China. China Environmental Science Press, Beijing (2009)
11. Ma, Z.: Environmental and resource economics [M]. Higher Education Press, Beijing (1999)
12. Hamilton, K.: Where is the wealth of nations? [M]. World Bank, (2005)
13. OECD: Towards green growth [R], 2011
14. UNEP: Towards a green economy: pathways to sustainable development and poverty eradication [R], (2011)
15. Li X., et al.: China green development index report 2010: provincial comparison [M]. Beijing: Beijing Normal University Publishing Group, (2010)
16. UNEP: Global Environment Outlook 5 (GEO5): Environment for the future we want [R], (2012)

# Chapter 2
# China-ASEAN Green Development Status

**Abstract** This chapter gives a detailed narrative and comparison of the green development status of China and ASEAN Member States in areas of green development history, social and economic development, industrial development transition, water and land resource conservation and utilization, air and water pollution treatment, solid waste management, environmental protection input and corresponding industrial development, energy use and climate change, ecosystem and biodiversity, as well as sustainable production and consumption. Given the data availability, the time frame of data for China is between 2001 and 2014, and between 2006 and 2013 for ASEAN. The comparative results are presented at the end, arguing that the level of green development process and priorities differs greatly between China and ASEAN, especially among ASEAN Member States. Economic development is relatively decoupled from resources and environment, but the challenge is still grim. Both China and ASEAN are faced with multifaceted challenge, including severe air and water pollution, growing energy use, biodiversity loss, and increasing waste discharge.

## 2.1 China

### *2.1.1 Review of the Green Development Course*

Aiming at balancing the relationship between the environment and economic growth, China's policy researchers and leadership have, in the past decade, advanced the understanding of and practices related to green development. To promote green development in China, they have introduced relevant laws, regulations, and policy instruments, which have shown outstanding effects and provided valuable experience. The transition of economic development model was first presented as a piece of key policy information in the Ninth Five-Year Plan for National Economic and Social Development and Vision 2010 released in 1996. Table 2.1 expresses the course of green economic development in China from 1996.



China-ASEAN Environmental Cooperation Center, *China-ASEAN Environment Outlook 1 (CAEO-1)*, https://doi.org/10.1007/978-981-10-6211-7_2

**Table 2.1** The course of green economic development in China (milestones upon 1996)

| Year | Keywords | Description |
|---|---|---|
| 1996 | 9th FYP for National Economic Development | "…emphasis of economic work should be attached practically to the transition of economic growth mode…" as stated clearly in the Ninth Five-Year Plan for National Economic and Social Development and Outline of Vision 2010 |
| 1997 | Energy Conservation Law | Implemented upon 1998, it aimed at promoting energy conservation nationwide, improving energy utilization efficiency and economic benefit, protecting the environment and securing national economic and social development by means of legislation |
| 2002 | Cleaner Production Promotion Law | Implemented upon January 1, 2003, it marked the start of the comprehensive promotion of cleaner production in China |
| 2003 | Scientific Outlook on Development | The Scientific Outlook on Development was presented as "sticking to putting people first, establishing a comprehensive, coordinated and sustainable development outlook and promoting the comprehensive development of economy, society and people" |
| 2005 | Renewable Energy Law | It marked that China resorts to legislation to promote and develop renewable energy, safeguard national energy security, prevent and control environmental pollution and ecological damage caused by energy utilization and accelerate the new energy revolution of renewable energy utilization |
| 2006 | Energy conservation and emission reduction | To reduce the energy consumption per unit of GDP by around 20% and total emission of major pollutants by 10% during the 11th Five-Year Plan |
| 2007 | Ecological civilization development | "Efforts should be made to construct ecological civilization, basically form an industrial structure, growth mode and consumption pattern where energy and resources are conserved and ecology and environment are protected" as specified in the report of the 17th National People's Congress |
| 2007 | Energy Conservation Law | Implemented upon 2008, it states clearly that "the central government should implement the basic national policy of resource conservation as well as the energy development strategy of promoting conservation and development simultaneously and putting conservation first" |
| 2008 | Circular Economy Promotion Law | Implemented upon 2009, it marked the start of the comprehensive promotion of circular economy in China |
| 2009 | $CO_2$ emission reduction | The Chinese government determined to incorporate reducing the $CO_2$ emission per unit of GDP by 40–45% by 2020, compared with 2005, as an |

(continued)

**Table 2.1** (continued)

| Year | Keywords | Description |
|---|---|---|
| | | obligatory target into the long-term planning of national economic and social development during and after the 12th Five-Year Plan |
| 2010 | 12th FYP for National Economic Development | The Outline of the Twelfth Five-Year Plan for National Economic and Social Development proposes "promoting green development and building a resource-saving and environment-friendly society" |
| 2010 | Revision of Renewable Energy Law | It presents regulations on the overall planning of the development and utilization of various kinds of renewable energy and establishes the important system of full-amount indemnificatory purchase |
| 2012 | Report of 18th CPC National Congress | The 18th CPC National Congress proposed "putting ecological civilization construction in a prominent position and integrating it into all aspects and overall processes of economic construction, political construction, cultural construction and social construction and striving to build a beautiful China and realize the sustainable development of the Chinese nation" |
| 2013 | 3rd Plenary Session of the 18th CPC Central Committee | The 3rd Plenary Session of the 18th CPC Central Committee said further that efforts should be made to improve the development achievements examination and assessment system, correct deviation that assesses official performance by economic growth rate alone, and increase the weight of resource consumption, environmental damage and ecological benefit indexes |
| 2013 | Air Pollution Prevention and Control Action Plan | The Plan states that over five years of efforts, the overall air quality nationwide should improve and days with heavy air pollution should decrease significantly; air quality of such regions as the Beijing-Tianjin-Hebei Region, Yangtze River Delta and Pearl River Delta should improve apparently. Every endeavor should be made to gradually eliminate heavy pollution weather and achieve an evident improvement in nationwide air quality with another five or more years |
| 2015 | Water Pollution Prevention and Control Action Plan | The Action Plan proposes that there shall be a periodical improvement of the water environment quality by 2020, dramatic reduction of seriously polluted water bodies, continuous improvement of drinking water guarantee, and strict control of groundwater over extraction |
| 2015 | 5th Plenary Session of the 18th CPC Central Committee | The 5th Plenary Session of the 18th CPC Central Committee put forward the development philosophy of being innovative, coordinated, green, open and shared |

(continued)

**Table 2.1** (continued)

| Year | Keywords | Description |
|---|---|---|
| 2016 | 13th FYP for National Economic Development | The 13th Five-Year Plan for National Economic and Social Development requires that during the 13th FYP period, China will focus on the improvement of environment quality, highlight solution to key ecological and environmental issues, beef up efforts to protect the ecology and environment, improve resource use efficiency, provide more high quality ecological products and work toward a strong and beautiful China with its people living in affluence |
| 2016 | Action Plan for the Prevention and Control Soil Pollution | By 2020, the exacerbating trend of soil pollution across the country shall be checked initially and the soil environment quality shall be overall stable |

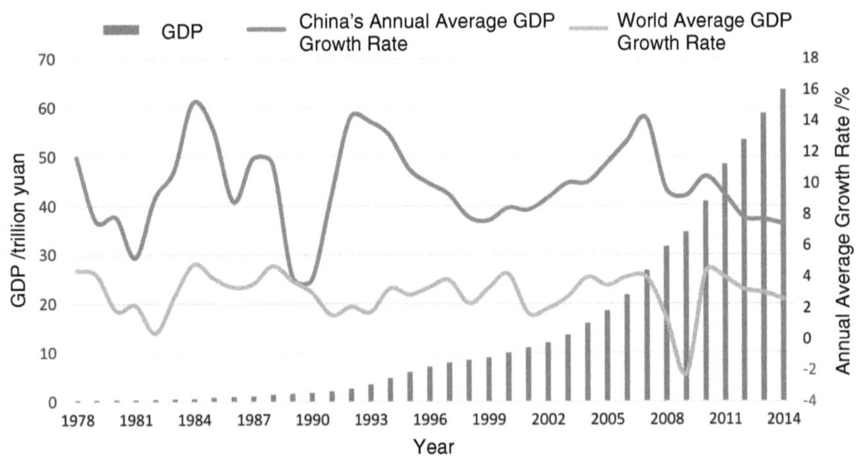


**Fig. 2.1** China's economic development upon the reform and opening-up. *Source* NBS Website, http://data.stats.gov.cn/ks.htm?cn=C01&zb=A0501; International Monetary Fund (IMF) Database (http://databank.worldbank.org/data/home.aspx). China's economic data all takes RMB as the monetary unit (Exchange rate: 1 RMB = 0.1496 USD)

## *2.1.2 Economic and Social Development*

### 2.1.2.1 Economic Growth

Since the founding of P.R. China, the country's economy has grown at an annual average growth rate of around 7.8%. China's annual average GDP growth has reached as high as 9.76%, especially over the 30 years of reform since 1978 (see Fig. 2.1). China's economy has maintained a high-speed growth for more than 30 years, much higher than the average GDP growth rate of the world. China is the largest exporter and manufacturer and the second biggest economy. While its economy develops rapidly, China's poverty rate has also dropped from 65% to less than 10%, with more than 500 million people lifted out of poverty [1].

#### 2.1.2.2 Income of Urban and Rural Residents

Along with the economic development, living quality indexes of urban and rural residents and their consumption standard have improved significantly (see Fig. 2.2). However, there is still a big gap between urban income and rural income. Since the reform and opening-up, the food expenses-to-consumption expenditure, i.e., Engel coefficient, of urban households has declined from 57.5% in 1978 to 35% in 2013, and that of rural households dropped from 67.7 to 37.7%. In 2012, rural Engel coefficient fell below the threshold of 40% for the first time [1].

### *2.1.3 Transition of Industrial Development*

#### 2.1.3.1 Adjustment of Industrial Structure

Over the past three decades, improvements in terms of the ratio of the three levels of industries with GDP have improved from 27.9:47.6:24.59 in 1978 to 9.2:42.7:48.1 in 2014 (see Fig. 2.3). The ratio of tertiary industry has apparently increased, the first industry tends to decline year on year while the secondary industry has witnessed fluctuations in its proportion and tended to decline after the 1980s, has started an evident rising tendency since the 1990s and registered a small and slowed growth in the first decade of this century [1].

Industry serves as the backbone of China's modernization as well as an important engine for its economic development. Jointly promoted by government policies and market forces, China, formerly with a poor industrial foundation and closed economy, has grown into an industrial power that leads industrial output of major products and plays a decisive role in the international market. The Made-in-China phenomenon has been globally recognized (see Table 2.2). At present, China maintains the economic growth driven by the secondary industry. Seen from economic structure, economic transition signifies the optimization of structure. Under the new normal status, more industrial attention is paid to activating new potentialities through technological transformation and speeding up structural adjustment [1].

#### 2.1.3.2 Strategic Emerging Industries

Issued in 2010, the Decision of the State Council on Accelerating the Fostering and Development of Strategic Emerging Industries (hereinafter referred to as the

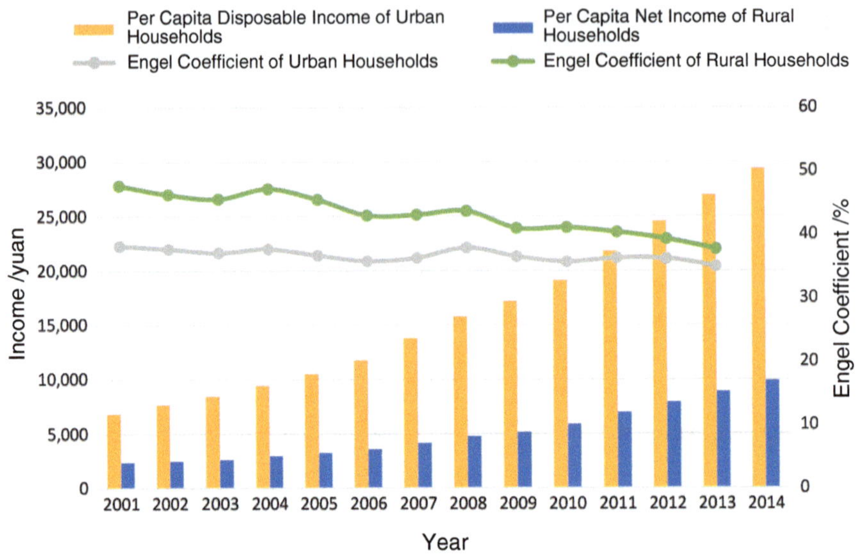


**Fig. 2.2** Yearly income and Engel coefficient of urban and rural residents. *Source* China Statistical Yearbook 2015. *Note* The Engel coefficient of resident households expresses the proportion of food expenses in the total consumption expenditure. UN establishes a standard for assessing the living standard of all countries based on the Engel coefficient, i.e. a country is evaluated as poor if its average household Engel coefficient is bigger than 60%; having enough to survive if its average household Engel coefficient is 50–60%; fairly well-off if its average household Engel coefficient is 40–50%; relatively affluent if its average household Engel coefficient is 30–40%; affluent if its average household Engel coefficient is 20–30%; and extremely affluent if its average household Engel coefficient is smaller than 20%

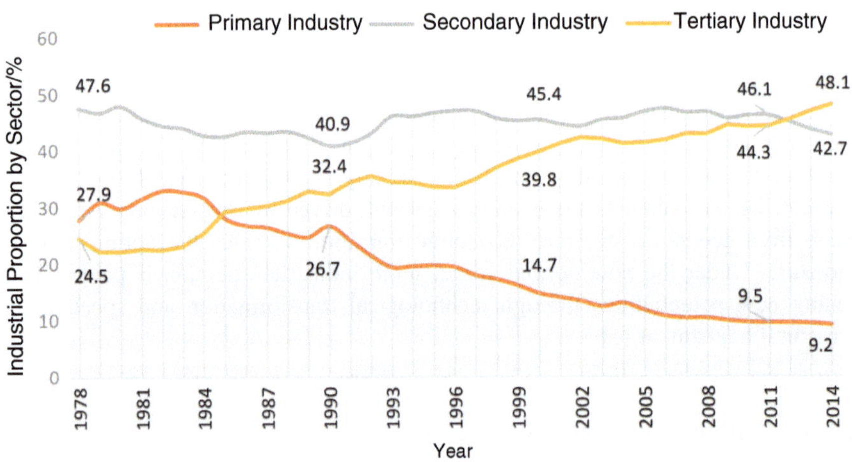


**Fig. 2.3** Changes in China's industrial structure, 1978–2014. *Source* China Statistical Yearbook 2015

**Table 2.2** Changes in China's leading industries, 2001–2014

| Rank | 2001 | | 2006 | | 2014 | |
|---|---|---|---|---|---|---|
| | Industry | Ratio (%) | Industry | Ratio (%) | Industry | Ratio (%) |
| 1 | Electric power and thermal power production and supply | 9.90 | Manufacturing of telecommunication equipment, computers and other electronic equipment | 8.24 | Manufacturing of telecommunication equipment, computers and other electronic equipment | 7.81 |
| 2 | Manufacturing of telecommunication equipment, computers and other electronic equipment | 7.47 | Ferrous metal smelting, rolling and processing | 8.14 | Transport and communication equipment manufacturing | 7.78 |
| 3 | Petroleum and natural gas exploitation | 7.41 | Electric power and thermal power production and supply | 8.04 | Manufacturing of chemical raw materials and chemicals | 7.54 |
| 4 | Transport and communication equipment manufacturing | 6.00 | Petroleum and natural gas exploitation | 6.96 | Ferrous metal smelting, rolling and processing | 6.50 |
| 5 | Manufacturing of chemical raw materials and chemicals | 5.88 | Manufacturing of chemical raw materials and chemicals | 6.28 | Electrical apparatus and equipment manufacturing | 6.13 |
| 6 | Ferrous metal smelting, rolling and processing | 5.62 | Transport and communication equipment manufacturing | 5.74 | Agricultural and sideline food processing | 5.82 |
| 7 | Textile | 5.09 | Electrical apparatus and equipment manufacturing | 5.37 | Nonmetallic mineral products | 5.33 |
| 8 | Electrical apparatus and equipment manufacturing | 5.06 | Textile | 4.61 | Electric power and thermal power production and supply | 5.17 |

Decision) specifies the emphasis for the development of strategic emerging industries at the present stage (see Table 2.3). At the end of 2010, added value of strategic emerging industries had accounted for 4% of GDP. In 2012, the State Council released the Twelfth National Five-Year Plan for the Development of Strategic Emerging Industries, promulgating a group of policy measures and indicating the direction for the fostering and development of strategic emerging industries. In 2014, China achieved a rapid development of strategic emerging industries, with the main business income and total profit of enterprises above the designated scale in the 18 key industries increased by 13.5 and 17.6%, respectively, evidently higher than the industrial average. Gradually, strategic emerging

**Table 2.3** Emphasis of the development of seven strategic emerging industries

| Industry | Key fields of development |
|---|---|
| New-generation information technology | New-generation mobile communications, integration of broadcast and television network, telecommunication network and the Internet, Internet of Things, cloud computing, integrated circuit, new-type display, high-end software, high-end server |
| Energy conservation and environmental protection | Energy-efficient, circular utilization of resources, advanced environmental protection technology |
| High-end equipment manufacturing | Aeronautic equipment, railway transportation equipment, ocean engineering equipment, high-end smart equipment |
| New energy | Nuclear energy, solar energy, wind energy, smart power grid, biomass energy |
| New material | New-type functional materials and structural materials, high- performance composite materials |
| Biology | Biological medicine, biological breeding, biological agriculture |
| New energy automobile | Plug-in hybrid electric vehicles and blade electric vehicles |

industries have serviced as an import force driving the adjustment of structure, the transition of mode and the benefiting of the people, and guided and driven, as an important role, economic and social development.[1]

## 2.1.4 *Conservation and Utilization of Resources*

### 2.1.4.1 Land Resources and Exploitation Intensity

China covers an area of 9.6 million km$^2$ of land and 4.73 million km$^2$ of waters. As a great power of land resources, the country's land area ranks third in the world. The complex topography and climate and the various types of land provide favorable conditions for the diversified operation and comprehensive development of agriculture, forestry, husbandry, sideline production, and fishery. In China, the utilization of land resources is quite complicated and confronted with such problems as the uneven distribution and the contradiction between protection and exploitation (see Fig. 2.4).

Since the founding of P.R. China, uncontrolled exploitation has been the norm and has not abated despite the introduction of sustainable development concepts. In recent years, China's national land development intensity has kept rising, with uneven fluctuations on the annual increment.

[1]National Development and Reform Commission (NDRC): Strategic Emerging Industries Developed Rapidly in 2014 in China, http://futures.hexun.com/2015-03-16/174085901.html.

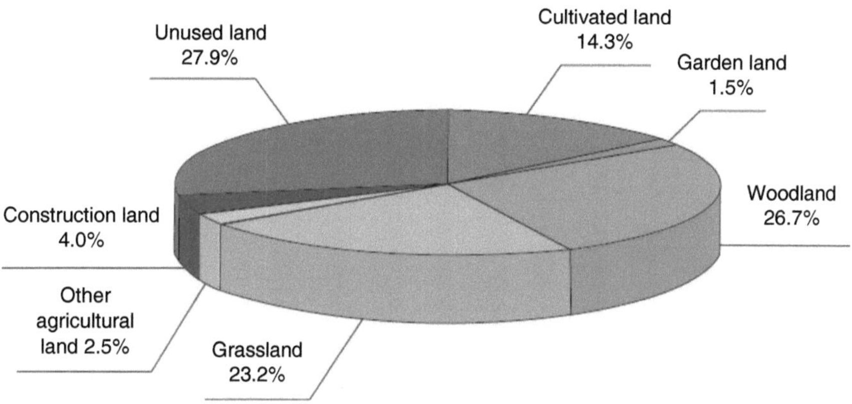


**Fig. 2.4** Land utilization in China, 2013. *Source* Communique on Land and Resources of China 2014

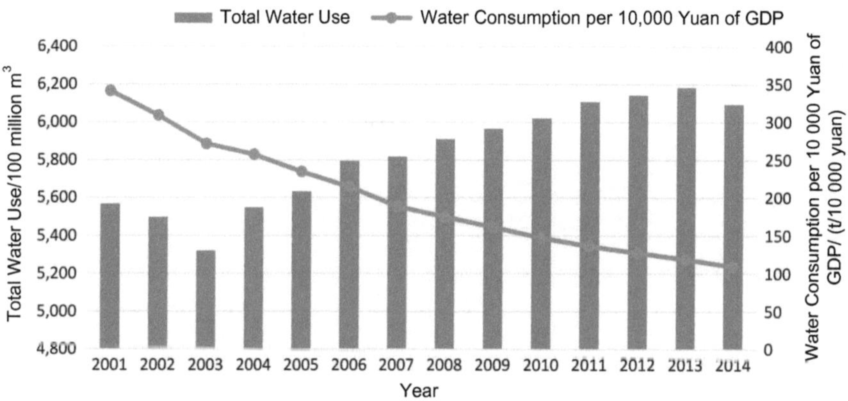


**Fig. 2.5** Changes in water-use efficiency in China, 2001–2014. *Source* China Statistical Yearbook 2015, at the prices of 2010

### 2.1.4.2 Water Resources and Water-Use Efficiency

In 2014, China's total water resources amounted to 2726.69 billion $m^3$, about 6% of the world's freshwater resources, ranking fourth. In average, China's precipitation and precipitation-converted water resources are lower than that of USA and European countries. Though among the best in terms of total water resources, China registers a small quantity of water resources per capita, 1998.6 $m^3$, merely 28% of the global average. Utilization of water resources tends to rise. In 2014, China's total water supply accounted for 22.4% of its total water resources for the year. Despite of the improvement (Fig. 2.5), the country's water-use efficiency remains lower than that of developed countries. Water resources are distributed unevenly and do not match the land resources and economic layout of China [2].

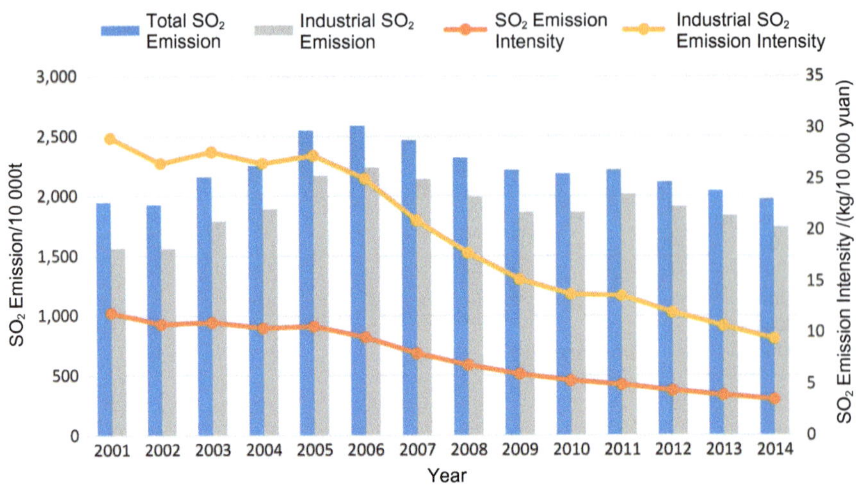


**Fig. 2.6** Analysis on $SO_2$ emission and emission intensity trends, 2001–2014. *Source* Data on pollutants are derived from Annual Statistic Report on Environment in China over years; GDP data are derived from China Statistical Yearbook 2013, at the prices of 2010

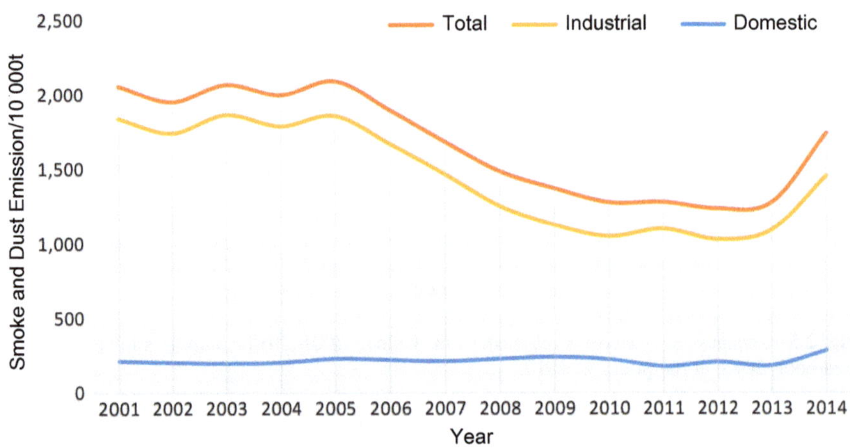


**Fig. 2.7** Smoke and dust emissions in China, 2001–2014. *Source* Annual Statistic Report on Environment in China 2001–2014

## 2.1.5 Environmental Protection

China's industrialization and urbanization have developed rapidly since the reform and opening-up, putting great strains on environmental protection. Hence, the contradiction between economic and social development and ecological and environmental protection has grown increasingly prominent. China has ranked top for the emissions of such major pollutants as COD, $SO_2$, $NO_x$, POPs, THP, EDS, and Hg [2].

#### 2.1.5.1 Atmospheric Environment

China's $SO_2$ emission peaked at 25.888 million tons in 2006, two times that of the USA. Although $SO_2$ emission intensity has been on a decline in the past decade (Fig. 2.6), it is still much higher than developed countries, even 5–10 times that of USA, U.K., and Japan, and higher than the global average. There's still a long way to control $SO_2$ emission, as the coal-focused energy consumption structure is hard to change and the space for the decline of emission intensity shrinks gradually. Industrial $SO_2$ removal rate has witnessed a quick rise, to 73% in 2014 [2].

China's total smoke and dust emission witnesses a small drop in fluctuations (see Fig. 2.7). From 2001 to 2013, total industrial smoke and dust emission tended to decline in fluctuations. In 2014, however, industrial and domestic smoke and dust largely increased. Along with the surging quantity of motor vehicles in China, highway traffic imposes a non-negligible pressure on air environment pollution. Industrial smoke and dust removal rate has remained high, to 98.2% in 2014.

In China, the emission of $NO_x$, a kind of major air pollutants, tends to keep rising (see Fig. 2.8). However, a sharp growth of 30% was witnessed in 2011, owing to the change of the specifications of environmental statistics. Industrial $NO_x$ removal rate is low, but keeps rising, reaching 27% in 2014, higher than the 19% in 2013 and 7.5% in 2012 [2].

Inhalable particles have been among the major air pollutants of China since the reform and opening-up. In the past decade, urban air quality has improved judging by normal pollutants. However, some cities are still bothered by heavy pollution (see Fig. 2.9). The pollution incurred by such composite pollutants as ozone and particulate matters grows increasingly serious, city clusters has witnessed compound pollution, and local rural areas registered a degradation of air quality. In 2014, air quality monitoring based on new standards was carried out in 161 cities at

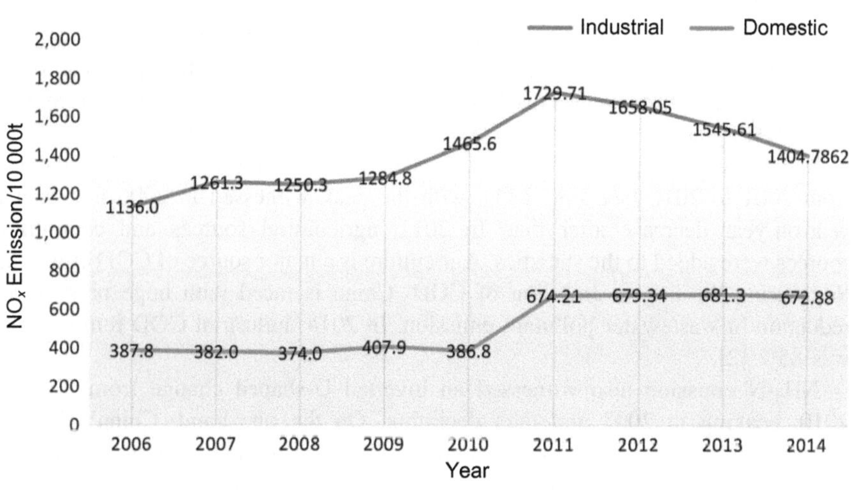


**Fig. 2.8** $NO_x$ emission in China, 2006–2014. *Source* Annual Statistic Report on Environment in China 2006–2014

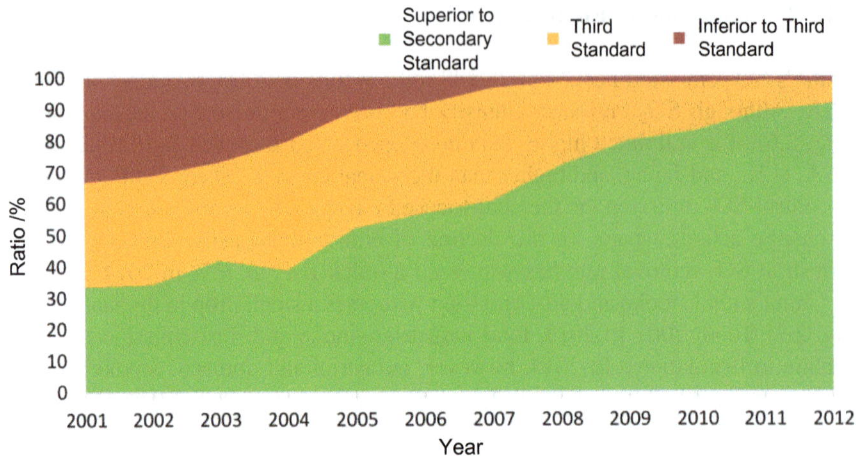


**Fig. 2.9** Changes in the ratios of cities by the level of air quality, 2001–2012. *Source* 2001–2012 Communique on Environmental Status in China, and Ambient Air Quality Standard (GB 3095–1996)

prefectural level and above, of which only 16 cities, 9.9% of the total, attained the standard of air quality, and 145 cities, 90.1% of the total, exceeded the standard [2].

#### 2.1.5.2 Water Environment

In China, total wastewater emission has increased year on year in the recent decade, of which the ratio of industrial wastewater emission has witnessed a year-on-year decline while domestic wastewater emission rate registered a year-on-year rise, from 52 to 71% in 2013, being a major source of wastewater emissions (see Fig. 2.10). From 2001 to 2014, wastewater emission intensity declined steadily by 52%; industrial wastewater emission intensity saw a more evident drop, from 37.5 tons/10,000 yuan to 11.1 tons/10,000 yuan, nearly a cut by 70% [2].

Main wastewater pollutants COD emission exhibited an inverted U-shaped curve from 2001 to 2010 (see Fig. 2.11), with the peak witnessed in 2006 and then a year-on-year decrease after that. In 2011, agricultural sources and centralized sources were added to the statistics. Agriculture is a major source of COD emission. Registering the largest emission of COD, China is faced with huge pressure of reduction in wastewater pollutant emission. In 2014, industrial COD removal rate was 85% [2].

$NH_3$-N emission also witnessed an inverted U-shaped change from 2001 to 2010, peaking in 2005 and then declining. On the one hand, China's $NH_3$-N emission is largely from domestic sources, whose proportion is on a year-on-year rise. On the other hand, its comprehensive $NH_3$-N emission intensity has also declined year by year in the recent decade. In 2011, agricultural sources and centralized sources were added to the statistics, with the former being the second

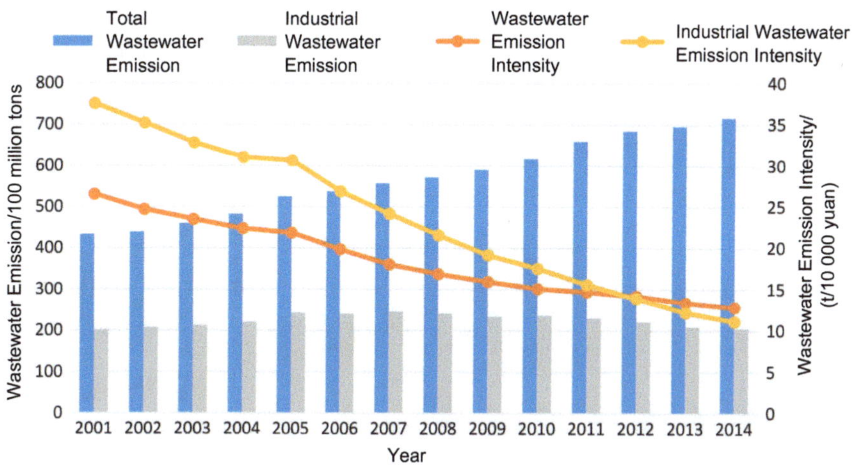


**Fig. 2.10** Wastewater and industrial wastewater emission and emission intensity in China, 2001–2014. *Source* Data on pollutants are derived from 2014 Annual Statistic Report on Environment in China; GDP data are derived from China Statistical Yearbook 2015, at the prices of 2010

largest source of $NH_3$-N emission and domestic sources ranking top. In 2012, county-level wastewater treatment rate approached 80%, but that of designated towns was smaller than 30%. Industrial $NH_3$-N removal rate was 82% in 2014 (see Fig. 2.12) [1, 2].

In the past decade, main heavy metal emission from industrial wastewater has been on a year-on-year decline in China (see Fig. 2.13). Industrial lead and arsenic emission accounts for the largest share, but is on a decline, from 80% in 2001 to 77% in 2014. Despite of the small proportion, industrial hexavalent chromium emission tends to rise, from 10% in 2001 to 15% in 2014. The proportion of industrial cadmium emission witnesses few changes, staying between 7 and 10% [2].

Water environment and pollution of the main river systems in China are under effective management. In 2014, state-controlled sections of surface water were mildly contaminated. Among China's seven most important river systems including the Yangtze River system, Yellow River system, Pearl River system, Huaihe River system, Liaohe River system, Haihe River system, and Songhua River system, the Haihe River system and the Liaohe River system have always been the highly contaminated ones in recent decades. In 2014, statistics showed that 96.2% of the water sources for centralized drinking water supply in 329 cities above the pre-fectural level met the standards. For surface water sources, total phosphorus, manganese, and iron exceeded the allowed limits, while iron, manganese, and $NH_3$-N exceeded the allowed limits for underground water sources (see Fig. 2.15) [2].

Rich in marine resources, China boasts a continental coastline as long as 18,000 km, a continental shelf covering over 200 $km^2$ and more than 6500 islands and administers a sea area of nearly 3 million $km^2$. 2014 Communique on Marine

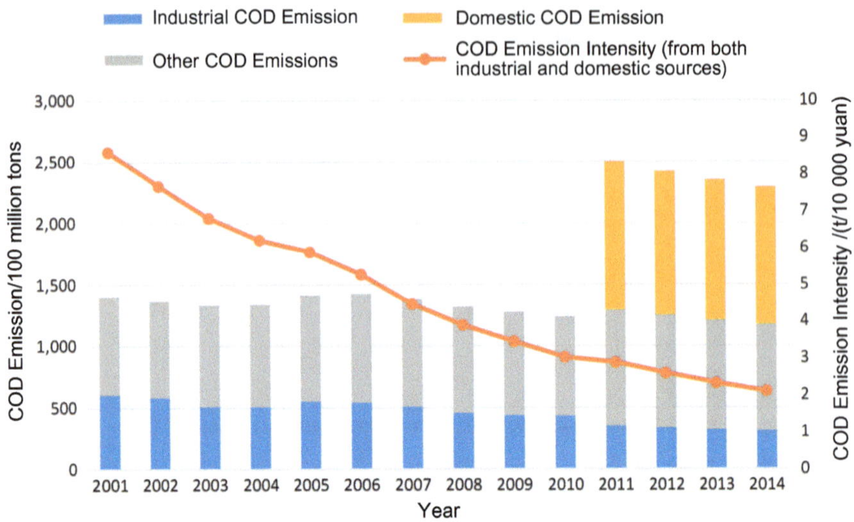


**Fig. 2.11** COD emission and emission intensity in China, 2001–2014. *Source* Data on pollutants are derived from 2014 Annual Statistic Report on Environment in China; GDP data are derived from China Statistical Yearbook 2015, at the prices of 2010

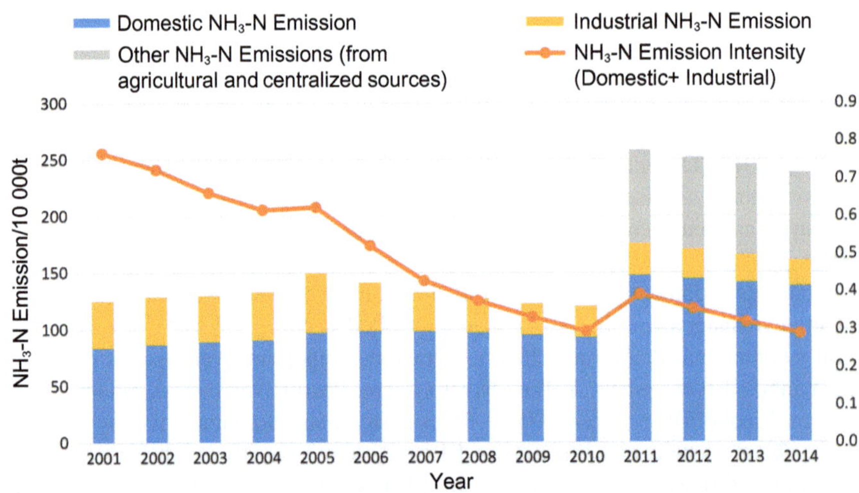


**Fig. 2.12** $NH_3$-N emission and emission intensity in China, 2001–2014. *Source* Data on pollutants are derived from 2014 Annual Statistic Report on Environment in China; GDP data are derived from China Statistical Yearbook 2015, at the prices of 2010

Environment Quality in China indicates that in 2014, the sea area under the jurisdiction of China was good on the whole, with the sea area up to Category I of seawater quality standard accounting for nearly 94%. Ocean sediments witnessed

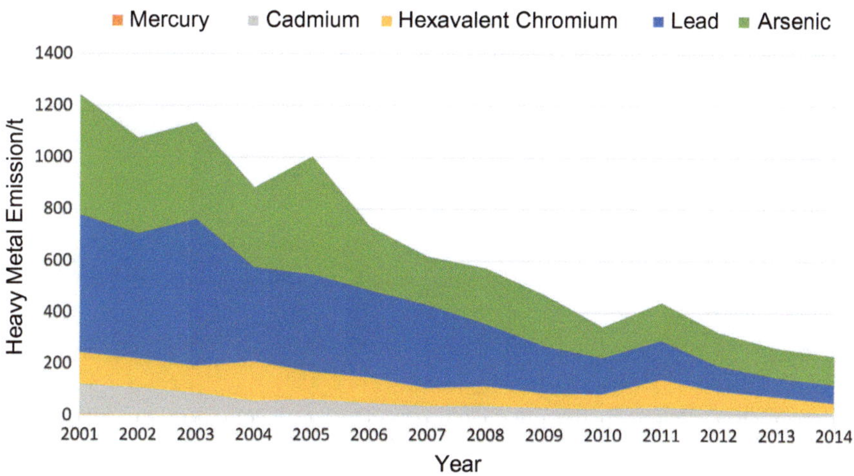


**Fig. 2.13** Changes in heavy metal emission from industrial wastewater in China, 2001–2014. *Source* Data on pollutants are derived from 2014 Annual Statistic Report on Environment in China

good quality. State-level marine conservation areas registered good quality on the whole. However, offshore areas were still bothered by such prominent problems as ocean water pollution, ecological damage, and frequent occurrence of disasters. For offshore areas, main pollutants are inorganic nitrogen, labile phosphate, and petroleum (Fig. 2.14).

#### 2.1.5.3 Solid Wastes

As industrial consumption of resources grows further in China, increasingly more solid wastes, mainly including gangue, coal ash, slag, residue, tailings, and hazardous wastes, are generated by the exploitation and utilization of resources (see Table 2.4). In addition, the influx of population in cities and the improvement of living standard also bring about serious discharge of household garbage, putting many cities under the "garbage siege" crisis. Owing to the development of the emerging electronic industry and the improvement of residents' consumption level, the generation of electronic wastes is accelerated by the upgrading of residents' electronic products, especially computers and mobile phones. China is still confronted with severe challenges in the level and mode of solid waste treatment due to the rapid increase of newly increased pollutants, though its innocent treatment of solid wastes and relevant plan and facilities are improving constantly [2].

Despite the year-on-year rise of industrial solid waste generation, the comprehensive utilization and treatment have kept rising, the emission has been on a

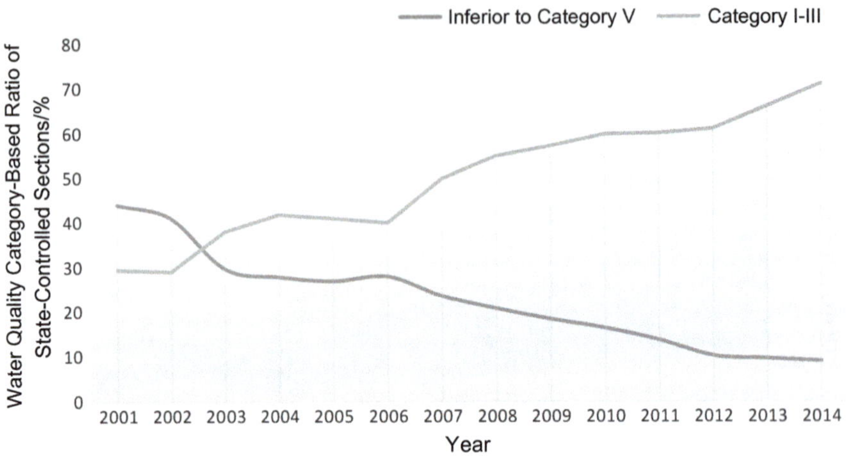


**Fig. 2.14** Changes in surface water quality in China, 2001–2014. *Source* 2001–2014 Communique on Environmental Status in China

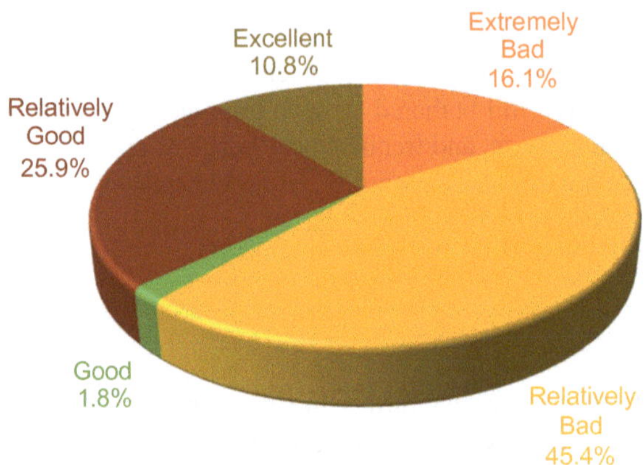


**Fig. 2.15** Underground water quality in China, 2014

decline, and the treatment rate has maintained at around 20% since 2001 (see Fig. 2.16).

#### 2.1.5.4 Environmental Protection Input

According to the experience of developed countries, a country's pollution won't be controlled effectively until its environmental protection input stays steadily between

**Table 2.4** Generation of four main kinds of solid wastes in China, 2006–2014

| | Generation of industrial solid wastes/100 million tons | Cleanup of urban household garbage/100 million tons | Generation of electronic wastes/10,000 tons | Urban sludge generation/10,000 tons |
|---|---|---|---|---|
| 2006 | 15.15 | 1.48 | 207.74 | 1104 |
| 2007 | 17.56 | 1.52 | 232.53 | 1517 |
| 2008 | 19.01 | 1.54 | 262.89 | 1452 |
| 2009 | 20.39 | 1.57 | 286.35 | 1655 |
| 2010 | 24.09 | 1.58 | 337.24 | 2116 |
| 2011 | 32.62 | 1.64 | 394.34 | 2268 |
| 2012 | 32.90 | 1.68 | – | 2418 |
| 2013 | 32.77 | 1.73 | – | 2635 |
| 2014 | 32.56 | 1.79 | – | 2801 |

*Source* 2001–2014 Communique on Environmental Status in China

1 and 1.5% of its GDP in a period of time when its economy grows at a high speed, and its environmental quality will improve evidently when that share reaches 3.0%. Since 2001, China's environmental protection input has kept increasing, at an annual average growth rate of 18%. In the meantime, the share of environmental protection input in GDP has been also on a rise, reaching 2.20% in 2014. During the 11th Five-Year Plan period, a rapider growth was witnessed, the entire inputs (including environmental investments and operation expenditure) totaling 2852.1 billion yuan (at the prices of current year) from 2006 to 2010. As a result, important material security has been provided for enhancing pollution mitigation, promoting environmental improvement, and relieving pollution trend (see Fig. 2.17). At the

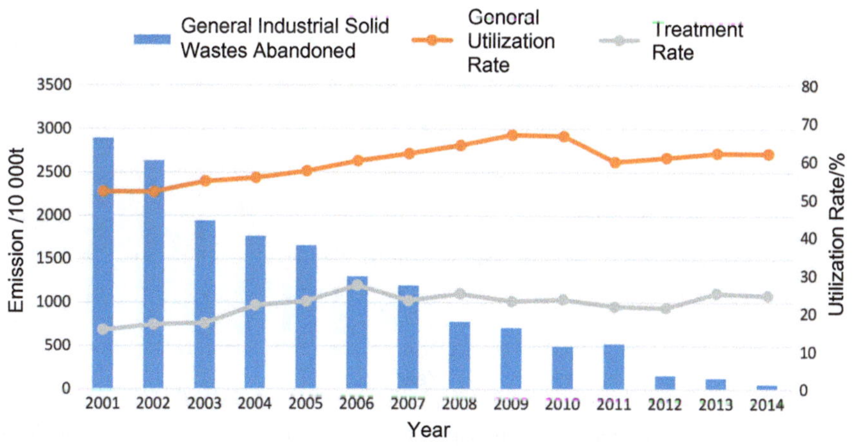


**Fig. 2.16** Emission, comprehensive utilization rate and treatment rate of industrial solid wastes in China, 2001–2014. *Source* 2001–2014 Communique on Environmental Status in China

earlier stage of the 12th Five-Year Plan (2011–2013), China's environmental protection input totaled 2.33 trillion yuan [2].

#### 2.1.5.5 Development of Environmental Protection Industry

From 2000 to 2008, China's annual average growth of environmental protection industry exceeded 20%, much higher than the GDP growth. In 2008, the gross income from environmental protection industry (including the comprehensive utilization of resources) amounted to 820 billion yuan, accounting for 2.76% of the GDP. Since the 11th Five-Year Plan period, China has created a huge demand for the environmental protection industry by promoting vigorously energy conservation and emission reduction, developing circular economy, and building a resource-saving and environment-friendly society. Hence, the environmental protection industry has grown rapidly and begun to take shape. Formerly focusing on the treatment of waste gas, wastewater, and solid wastes in the early stage, it has been a multi-industry, trans-regional system covering environmental protection products, environmental service, clean products, recycling of wastes, and ecological protection. In 2011, environmental protection-related industries involved 23,820 units and 3.195 million employees and obtained an annual operating revenue of 3075.25 billion yuan, an annual operating profit of 277.72 billion yuan and an annual contract amount of export of 33.38 billion US dollars. A full-category

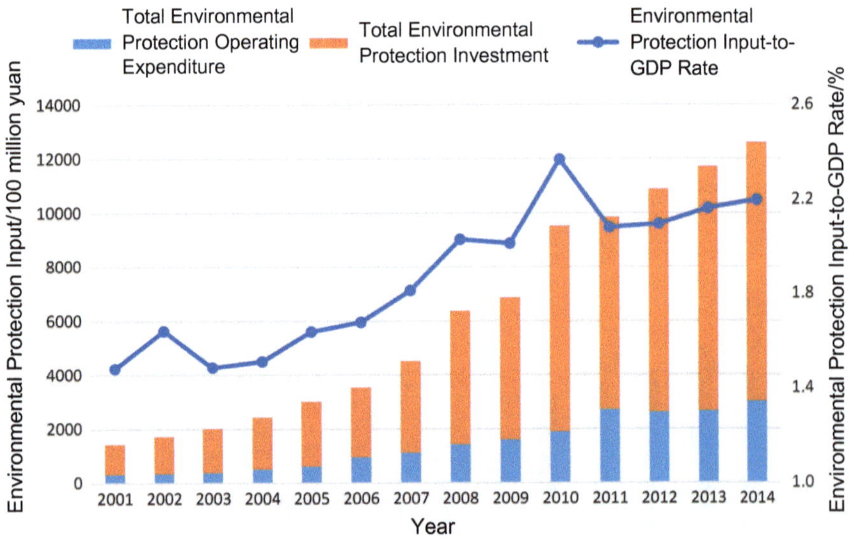


**Fig. 2.17** Total environmental protection input and its proportion in GDP of China over Years. *Note* Environmental protection input consists mainly of environmental protection operating expenditure and environmental protection investment; relevant data are derived from 2001–2014 Annual Statistic Report on Environment in China and China Statistical Yearbook 2002–2015

industrial system has taken the initial shape along with the constant extension of its industrial field, the rapid upgrading of technology and equipment, the continuous increase of product categories, and the evident improvement of service.

## 2.1.6 Energy and Climate Change

### 2.1.6.1 Energy Consumption

Along with the speeding-up of industrialization and urbanization, China's energy consumption has witnessed a rapid growth since the reform and opening-up. The energy consumption was merely 603 million tons of standard coal in 1980 and increased to 1.47 billion tons of standard coal in 2000, at an annual average rate of 4.6% (see Fig. 2.18). In 2001, as China joined WTO, its national economy ushered in a new stage of development. With a surging increase, the energy consumption had totaled 4.26 billion tons of standard coal in 2014, up 190% than 2000 and at an annual average rate of 7.9%.

Coal has always accounted for a big share, around 70%, of the primary energy consumption structure in China. As the relatively clean natural gas and other energy resources (hydropower, nuclear power, wind power, etc.) are applied gradually, coal consumption-to-total energy consumption rate has dropped, to 66.0% in 2014, while natural gas and other energy resources (hydropower, nuclear power, wind power, etc.) have registered a bigger proportion, 16.9% in 2014. Compared with developed countries, China witnesses an excessively small ratio of petroleum and natural gas in its energy consumption and the coal-focused energy consumption structure has intensified the need for environmental and ecological protection (see Fig. 2.19).

Seen from the intensity of energy consumption, per capita energy consumption has been rising since the reform and opening-up, along with the increasing improvement of living standard (see Fig. 2.20). After 2001, per capita energy consumption quickened its growth, with a rise of 169% from 2000 to 2014. On the other hand, China's energy consumption per unit of GDP is on a decline, as industrialization and urbanization keep speeding up and industrial technology and energy utilization efficiency improve constantly [1].

### 2.1.6.2 Climate Change

Climate warming in China is consistent with the global tendency. According to the monitoring results released by China Meteorological Administration, China witnessed a rise of 1.1 in its surface temperature in the past century (1908–2007) and experienced 21 mild winters since 1986, with 2007 being the warmest year since 1961 under systematic meteorological observation. Based on the global $CO_2$ emission tendency report 2015, China's $CO_2$ emission has been on a rise, but at a

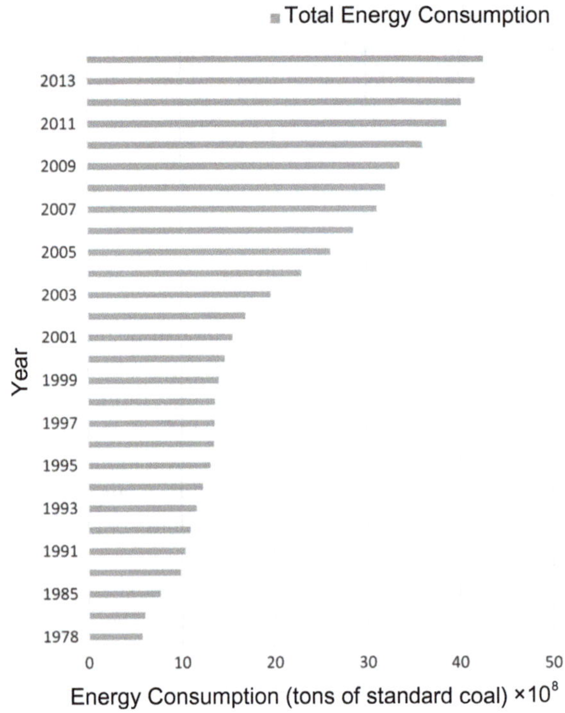


**Fig. 2.18** Total Energy Consumption, 1978–2014. *Source* China Statistical Yearbook 2015

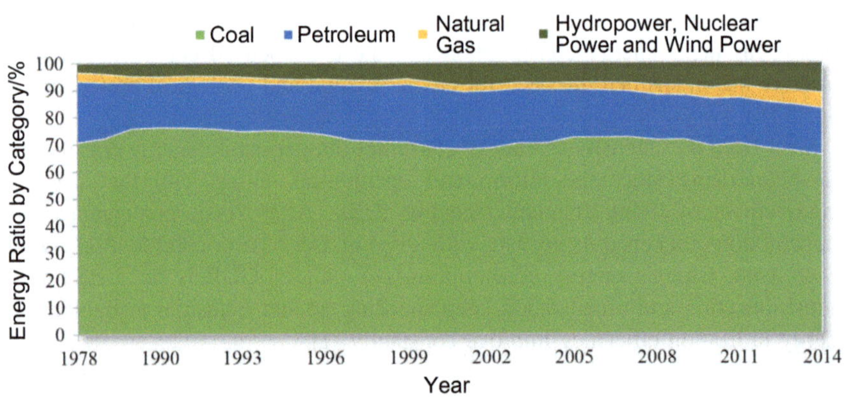


**Fig. 2.19** Energy Consumption Structure, 1978–2014

smaller rate; per capita $CO_2$ emission has increased year on year and the increase quickened after 2002 and slowed down after 2011; China's carbon emission intensity has declined remarkably, at a bigger speed than other countries (see Figs. 2.21 and 2.22) [3].

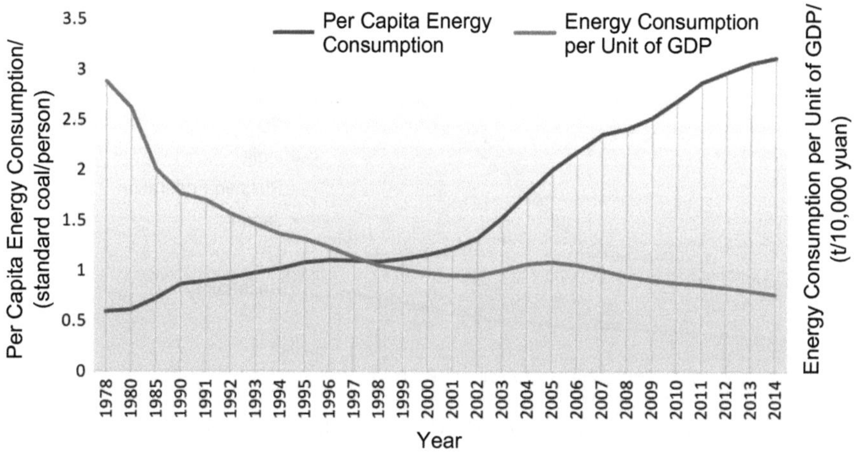


**Fig. 2.20** Energy consumption per unit of GDP and per capita energy consumption, 1980–2014. *Source* China Statistical Yearbook 2015

China has always been very active in promoting global climate control. In November 2009, China announced the action plan of reducing its $CO_2$ emission per unit GDP by 40–45% in 2020 relative to 2005 levels and incorporated it into its mid- and long-term planning for national economic and social development as a binding indicator. In June 2015, China submitted its Intended Nationally Determined Contribution to address climate change to the Secretariat of the United Nations Framework Convention on Climate Change, pledging to reduce its $CO_2$ emission per unit GDP by 60–65% in 2030 relative to 2005 levels. This is not only the required work of China as a contracting party, but also the biggest effort China can make to hit the target of the convention. The report of 18 non-governmental organizations including World Wide Fund said that China's climate action target has exceeded its "fair share."

## 2.1.7 Ecosystem and Biodiversity

### 2.1.7.1 Natural Reserves and Biodiversity

By the end of 2014, China has established 2792 natural reserves, covering an area of 146.992 million ha, 14.8% of national territorial area, and 12% higher than the world average. Among them are 407 state-level natural reserves, covering an area of 96.516 million ha. A local protection network of biodiversity, centering on natural reserves, has taken shape basically (see Fig. 2.23) [1, 4].

Covering a vast area, China possesses complex climate and topography, a rich variety of endemic species, anciently rooted floras and complicated spatial

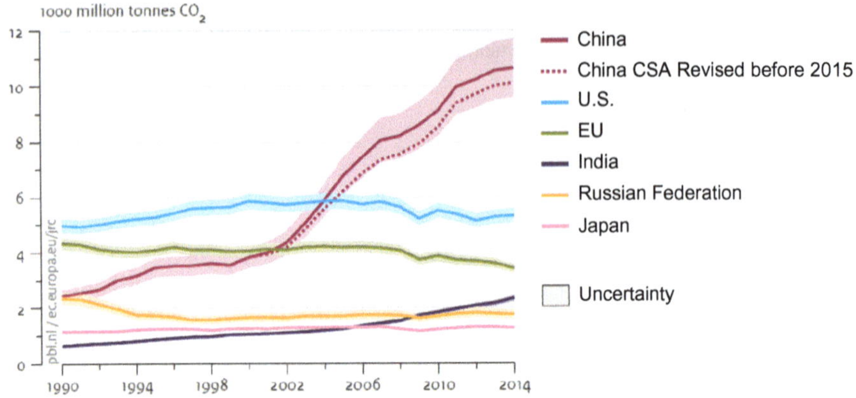


**Fig. 2.21** Great emission countries and EU $CO_2$ emission tendency. *Source of picture* Trends in Global $CO_2$ Emissions: 2015 Report

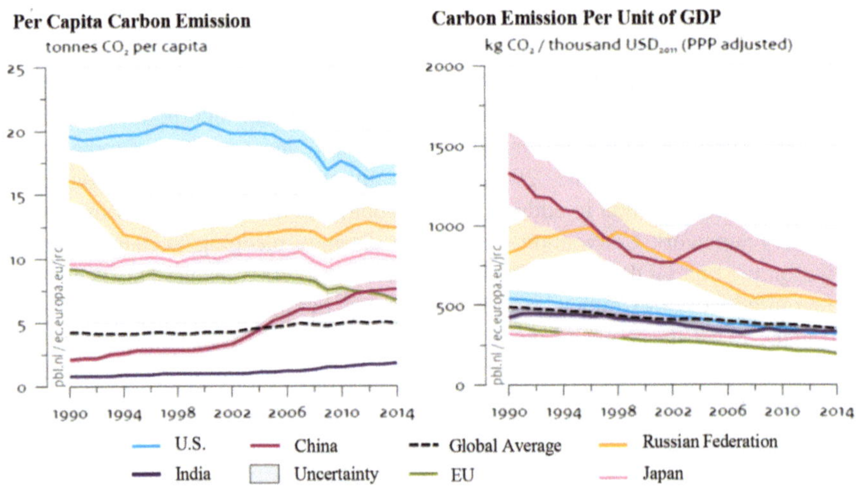


**Fig. 2.22** Great emission countries and trends in EU per capita $CO_2$ emission and $CO_2$ emission per unit of GDP. *Source of pictures* Trends in Global $CO_2$ Emissions: 2015 Report

distribution of biodiversity. It is one of the 12 mega-diversity countries, the 8 centers of origin of crops, and the 4 centers of origin of cultivated plants. The Chinese government and relevant sectoral administrations have implemented a group of plans and programs in the protection of natural reserve, wetland, aquatic organism, as well as livestock and poultry genetic resources and combined the development of biotechnology and the promotion of the sustainable utilization of biological resources. China's 5th national report submitted in March 2014 on the

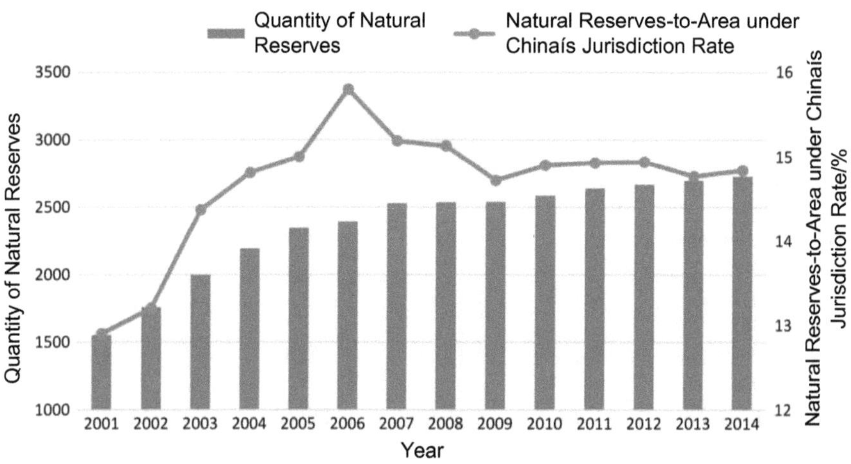


**Fig. 2.23** Quantity and area of natural reserves in China. *Source* China Statistical Yearbook 2002–2015

implementation of Convention on Biological Diversity summarized the status quo and trend of China's biodiversity.

The endangered rate (critically endangered, endangered, and vulnerable animals) of invertebrate animals is 34.7% and vertebrate 35.9%. China currently has 3767 endangered plants, accounting for 10.9% of the higher plants of the country. The total number of higher plants which need attention and protection reached 10,102, accounting for 29.3% of the total. China's genetic resource loss is very serious. Results of the second national survey for husbandry genetic resources show that 15 local varieties disappeared and the number of local varieties above general level is decreasing.

The pressure for biodiversity conversation mainly comes from fast population growth and accelerated industrialization and urbanization which lead to the degeneration of wild species and loss of their habitats. Other threats are over development of natural resources, environmental pollution, large-scale plantation of single species and invaded alien species, and climate change.[2]

#### 2.1.7.2 Protection of Forest and Wetland

Forest, grassland, and wetland play a significant role in improving environmental quality, regulating air temperature, preventing wind, fixing sand, protecting water and soil, and maintaining ecological balance. According to the 8th national forest resources inventory (2009–2013), forests cover an area of 208 million ha

[2]CAEC, Implementation Progress of China-ASEAN Biodiversity Strategy and Action Plan, China Environmental Science Press, 2016.

**Table 2.5** Forest resources inventory of China over years

| Inventory and its time | Forest area/10,000 $km^2$ | Forest coverage rate/% | Forest stock/100 M $m^3$ | Standing forest stock/100 M $m^3$ |
|---|---|---|---|---|
| 1949 | 76.0 | 7.9 | | |
| The first inventory (1973–1976) | 122.0 | 12.7 | 86.56 | 95.3 |
| The second inventory (1977–1981) | 115.2 | 12.0 | 90.28 | 102.6 |
| The third inventory (1984–1988) | 124.6 | 12.98 | 91.41 | 105.7 |
| The fourth inventory (1989–1993) | 133.6 | 13.92 | 101.37 | 117.85 |
| The fifth inventory (1994–1998) | 158.94 | 16.55 | 112.67 | 124.88 |
| The sixth inventory (1999–2003) | 174.91 | 18.21 | 124.56 | 136.18 |
| The seventh inventory (2004–2008) | 195.6 | 20.36 | 137.21 | 149.13 |
| The eighth inventory (2009–2013) | 208 | 21.63 | 151.37 | 164.3 |

nationwide, with a forest coverage rate of 21.63% (see Table 2.5). China ranks fifth for forest area, sixth for forest stock, and still top for the area of planted forest. In China, forest resources witness a constant increase in quantity, a steady improvement in quality, and a continuous rise of efficiency. While global forest resources decrease on the whole, China witnesses a constant growth and registers the biggest speed of growth [5, 6].

The forest ecosystem is fragile. China's forest coverage rate remains much lower than the world average, 31%. Moreover, forest quality is poor, with a forest growing stock of 89 $m^3$ per ha, while the number is 300–320 $m^3$ per ha or even higher in some developed EU countries. Besides, China's per capita forest resources available are merely one-sixth of the global average, and its per capita grassland available is only half the world average. Grassland quality declines gradually, and grassland ecosystem is damaged severely due to the long-term irrational utilization and development of land. At present, desertification land accounts for 18% of China's national territorial area (see Fig. 2.24). Area of soil erosion is about 30.7%.

The first national wetland resources inventory (1995–2003) indicates China's wetland covers an area of 38.48 million ha, accounting for 4.01% of its national territorial area, for which China ranks top in Asia and fourth in the world. With a rich biodiversity of wetland, China possesses as many as 271 species of waterfowl and is an important area for the world's protection of rare and endangered

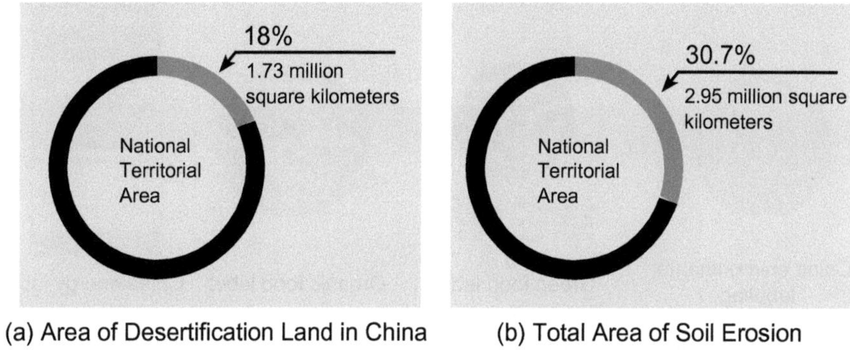


**Fig. 2.24** Land resources damage in China. *Source* China Statistical Yearbook 2015

waterfowl. Up to now, wetland damage remains hard to contain. In the recent 50 years, China's wetland damage rate has reached 21.6%, according to long-term researches and statistics.

## *2.1.8 Sustainable Production and Consumption*

### 2.1.8.1 Eco-Label

In China, eco-labeling is referred to as "environmental label system." In August 1993, the Chinese government confirmed officially that the environmental label should be composed of a mountain, water, a sun and ten rings, namely the "ten-ring label." Up to now, China has built up the environmental labeling system for agricultural products, represented by green food mark and organic food mark; the environmental labeling system for manufacturing and building products, represented by China environmental level; and the energy efficiency labeling system, represented by the energy efficiency label (see Fig. 2.25).

### 2.1.8.2 Environmental Awareness

As demonstrated by the issuance of the country's first National Ecological Civilization Awareness Survey and Research Report in 2014, China witnesses "high recognition, poor knowledge and insufficient practice" in the public awareness of ecological civilization, with the public recognition, knowledge, and practice rates being 74.8, 48.2, and 60.1%, respectively (see Fig. 2.26). As public awareness of ecological civilization features high "dependence on the government," the respondents agree widely that government and environmental departments should be solely responsible for developing ecological civilization. The report finds out

**Fig. 2.25** China environmental labels

that people's economic status and educational background greatly affect their awareness and knowledge of ecological civilization. Moreover, the respondents are quite worried about the current status of environment. Their biggest concerns include haze, drinking water safety, heavy metal pollution, etc.

#### 2.1.8.3 Modes of Life and Consumption

China is among the largest and most potential consumption markets. It remains at the stage of development, with a lower per capita consumption capacity. In China, the largest pressure on sustainable consumption is derived from the increasing discharge of wastes. In recent years, China has witnessed a rapid decline in green travel sharing rate, especially for bicycles and walk, with a decrease of even above 10% in many cities in the past decade.[3] In the recent five years, growth of cars has exceeded 20% in many cities. Presently, government departments and officials are less capable of understanding and using new media to satisfy the realistic demand, though new media has been an important way for the general public to acquire environmental information, express their will, participate in environmental decision making, supervise environmental management and law enforcement, and express their green selections. Along with the gradual improvement of living standard, people have raised new appeals for environmental quality, so efforts should be made to seek and establish resource-saving and environment-friendly modes of life and consumption.

[3]Source: CCICED Special Policy Study on "Promoting Urban Green Travel".

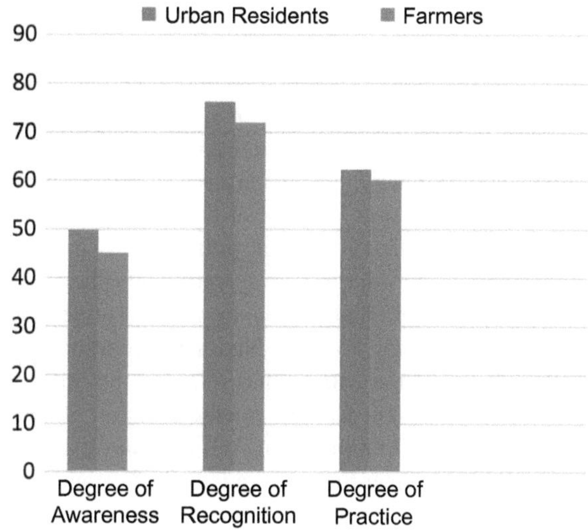


**Fig. 2.26** Comparison between urban residents and rural residents by scores of ecological civilization awareness. *Source* National Ecological Civilization Awareness Survey and Research Report

## 2.2 ASEAN

### *2.2.1 Review of ASEAN Green Development Course*

ASEAN is rich in natural resources and among the most populous regions in the world. Its economy depends largely on industry, service, and agriculture, of which industry and service register the largest contribution to GDP. The ASEAN suffers from impacts of unsustainable development, over exploitation of natural resources, and the degradation of environmental and ecosystem services. The ASEAN economies are fully aware of these threats and impacts brought about by traditional development and have thus mobilized the transition toward sustainable industrial development.

In the 1990s, the Environmental Impact Agency of Indonesia issued Indonesia Environmental Protection Strategy Plan (1994–1998). In 2007, the Indonesian government formulated a "national action plan" specific to the fight against climate change. At present, the country is striving to establish a low-carbon economy. Singapore is rich in the experience of practice. The government developed the Singapore Green Plan 2012 (SGP 2012) in 2001 and has formulated Sustainable Singapore Blueprint 2015. In Brunei, environmental protection strategy was included into the eight strategies in the "2035 long-term outlook" as specified in the Long-term Development Plan of Negara Brunei Darussalam released in 2008. In 2009, Myanmar started to implement the sustainable development strategy, which aims to realize the happiness and joy of the Burmese people and sets up three objectives for green development transition from the perspectives of environment, economy, and society, including the sustainable management of natural resources, the comprehensive development of economy, and the sustainable development of society. In 2010, Malaysia promulgated the 10th national development plan, presenting "ten concepts," including "cherishing natural resources and environment" and "five strategies," of which an important one is building a good environment and improving living quality. In 2013, to realize green development, relevant departments of Cambodia formulated and implemented the 2013–2030 Green Growth Framework Strategy, which ensures a balance between economic growth, environmental protection, and social development while promoting economic growth, through standardizing national policies and strategic development plans for green development, in order to help government officials with the management of natural resources. As for Thailand, environmental protection is among the 6 key points of the 11th Five-Year Plan for National Economic and Social Development (2012–2016). Besides, the country has released the Strategic Plan of Green Growth. The Laotian government has presented the vision of "clean, green and beautiful cities" in its National Environmental Action Plan (2011–2015).

ASEAN Member States have launched cleaner production plan in manufacturing successively. For instance, Malaysia has released a roadmap of cleaner production; Vietnam has promulgated a national clean production action plan and in 1998 founded the National Clean Production Center; the Thailand government issued the National Clean Production Development Plan (2002–2011) in January 2002; the government of the Philippines unveiled the Philippines Environmental Partnership Plan and estimated the implementation of pollution management projects by private enterprises so as to attain environmental standards better. The Singapore government has established an energy efficiency improvement aid program to provide 50% of the capital needed by energy audit, encourage enterprises to calculate their energy consumption, and tap the potentialities of energy efficiency improvement. Other ASEAN Member States have also implemented the environmental label plan to encourage sustainable development. For instance, the Singapore Green Label Scheme (SGLS) was launched in 1992; Thailand Business Council for Sustainable Development initiated the country's green label plan in 1993; Indonesia launched the ecological label authentication and permission plan on the World Environment Day of 2004; and Vietnam approved its national ecological label plan in 2009.

ASEAN also makes efforts in various respects to fulfill its promise to promote green development at regional level and has promulgated a great many declarations and proposals relating to environment since it placed environmental issues on the agenda over 30 years ago. Presently, ASEAN plays an important role in promoting the sustainable development of regional environment and the implementation of multilateral environmental agreements. In 1977, under UNEP promotion and support, ASEAN formulated the first ASEAN Sub-regional Environment Plan [ASEP-I (1978–1982)], including 6 prior fields and more than 100 environmental projects implemented by ASEAN Expert Group on the Environment (AEGE) under ASEAN Science and Technology Committee. In 1981, ASEAN established the ministerial meeting mechanism on environment. In the Manila Declaration on the ASEAN Environment in 1981, the objective is defined as "to ensure the protection of the ASEAN environment and the sustainability of its natural resources so that it can sustain continued development with the aim of eradicating poverty and attaining the highest possible quality life for the people of the ASEAN countries."[4] Subsequently, ASEAN formulated and implemented the 2nd and the 3rd subregional environment plan, laying a foundation for the subregional environmental cooperation mechanism of ASEAN. After 1989, ASEAN Summit and ASEAN Ministerial Meeting on Environment were established. In 1992, ASEAN Summit was held in Singapore, when ASEAN Member States reached agreed unanimously with environmental issues and sustainable development and ASEAN Member States promised to respond positively to the Agenda 21. At the ASEAN 2020 Vision Forum held in Kuala Lumpur in 1997, ASEAN heads approved the ASEAN 2020 Long-term Objectives. In 1998, Hanoi Action Plan (1999–2004), namely ASEAN Strategic Action Plan on Environment (1999–2004), was approved in Hanoi. In November 2002, the Framework Agreement on Comprehensive Economic Cooperation between the People's Republic of China and the Association of Southeast Asian Nations was signed at the 6th China-ASEAN Summit. In 2007, the 13th ASEAN Summit approved unanimously the ASEAN Socio-Cultural Community Blueprint 2009–2015. In 2009, the Ministry of Environmental Protection of P.R. China and ASEAN Member States formulated jointly the China-ASEAN Environmental Protection Cooperation Strategy (2009–2015). In the Manila Declaration on Green Industry, main developing economies of ASEAN and other Asian countries promised jointly to establish a framework of policies, monitoring, and regulations so as to promote the low-carbon and efficient transition of industry. In 2009, in particular, the ASEAN Day took "green development" as the theme, demonstrating ASEAN's firm commitment to exercising sustainable development. In 2010, China-ASEAN Environmental Cooperation Center (CAEC) was founded.

Environmental protection is presented among the four important challenges in the ASEAN 2030 published in 2014 [7, 8].

At present, the important action plans for the implementation of ASEAN-China, Japan, and South Korea leadership project on sustainable production and consumption, ASEAN ecological school project, and ASEAN sustainable city plan have shown ASEAN's solid commitment to promoting green development. ASEAN Member States and other stakeholders have facilitated further regional green development through multilateral coordination, cooperation, and exchange. Presently, ASEAN is striving for the objective of building up the ASEAN Community by 2015.

[4]Source: Manila Declaration on the ASEAN Enviroment, http://enviroment.asean.org/manila-declaration-on-the-asean-enviroment/.

### 2.2.2 Economic and Social Development

ASEAN economies differ in terms of scale (gross domestic product, GDP). For instance, Indonesia was the largest economy and Singapore the smallest one in 2013, which is closely related to population size. Indonesia registers the largest population in ASEAN and Brunei the smallest. In 2013, Singapore, Brunei, and Malaysia were in the development leaders, with the per capita GDP being 10,000 US dollars or more, which was 4000–8000 US dollars for other member states. In recent years, ASEAN Member States have experienced a steady economic growth (see Fig. 2.27).

In 2014, ASEAN Member States witnessed a steady development of economy on the whole. In the recent decade, ASEAN has achieved an average annual growth rate of 5–7%. Laos witnessed the fastest growth in 2006–2013, with an average growth rate of 9.3%, and Brunei was on the contrary, with a negative growth of economic aggregate in 2013 (see Fig. 2.28) [9].

Seen from the industrial structure, agriculture accounts for a bigger proportion, over 20%, in Cambodia, Laos, and Myanmar than that of other ASEAN Member States. The rate reached 17.5% in Vietnam in 2013 and is nearly zero in Singapore and Brunei. Thailand and Brunei register a higher rate of industry, around 50%, and service accounts for over 50% in Malaysia and Philippines (see Fig. 2.29).

ASEAN suffers a serious explosion of population. The most populated countries have a high speed of growth. Distribution of population is uneven, and a large number of population are located in coastal areas. Few reside in inland mountainous areas, and there is an imbalance between males and females.

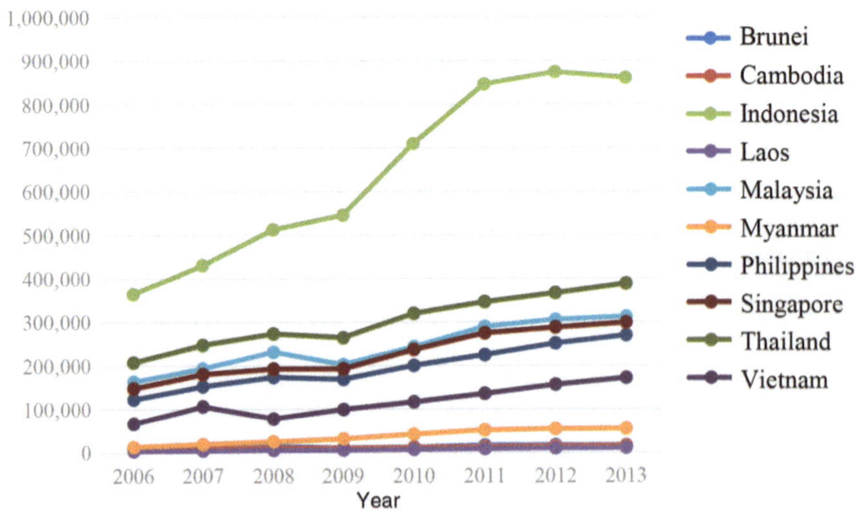


**Fig. 2.27** Economic aggregate of ASEAN Member States. GDP at Current Prices/Million US Dollars. *Source* ASEAN Statistical Yearbook 2014

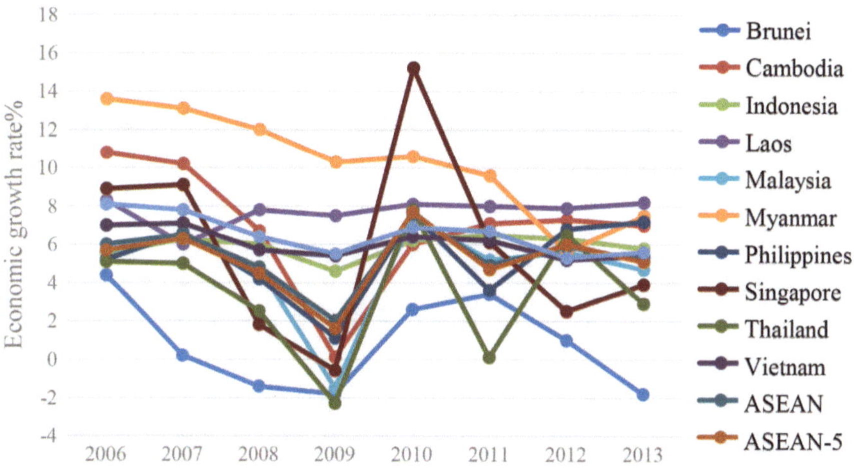


**Fig. 2.28** Economic growth of ASEAN Member States. *Note* ASEAN GDP growth rate is figured out by the weighted average of the PPP-GDP rate adopted in October, 2014 of the IMF-WEO database. ASEAN-5 includes Indonesia, Malaysia, the Philippines, Singapore and Thailand. BCLMV represents Brunei Darussalam, Cambodia, Lao People's Democratic Republic, Myanmar and Vietnam. *Source* ASEAN statistical Yearbook 2014

## 2.2.3 Conservation and Utilization of Resources

### 2.2.3.1 Land Resources

Possessing rich land resources, ASEAN covers an area of more than 4.43 million km$^2$, about 3% of the global land area. Among ASEAN Member States, Indonesia registers the largest national territorial area, 1.86 million km$^2$, and Singapore the smallest one, 710 km$^2$. Agricultural land in ASEAN amounts to 117,272,000 ha, about 27.1% of its land area, of which the Philippines, Thailand, Vietnam, and Cambodia account for more than 30%, with the Philippines and Thailand ranking top by 39.6 and 38.5%, respectively. The agricultural land-to-land area rate is lower than 10% for Brunei and Laos and merely 1% for Singapore.

Seen from the protection of land resources,[5] 13.2% of ASEAN land is under protection (see Fig. 2.30), of which Brunei accounts for 49.77%, the largest one, Cambodia 26.65%, Thailand 20.36%, and Malaysia 18.28%, and it's lower than 10% for Vietnam, Singapore, and Myanmar, and merely 4.57% for Singapore. Land under protection accounts for only 14.89% of the national territorial area in Indonesia, but 42% of the ASEAN area [10].

Despite the imbalance in the area of land protected, ASEAN Member States have witnessed a growth in the recent decade. Additional land has been protected,

[5]Source: Peng Bin, Liu Xiaoxue, Yang Zhenzhong, ASEAN State of Resources and Environment and Potential of Cooperation, Beijing, Social Sciences Academic Press, 2013.

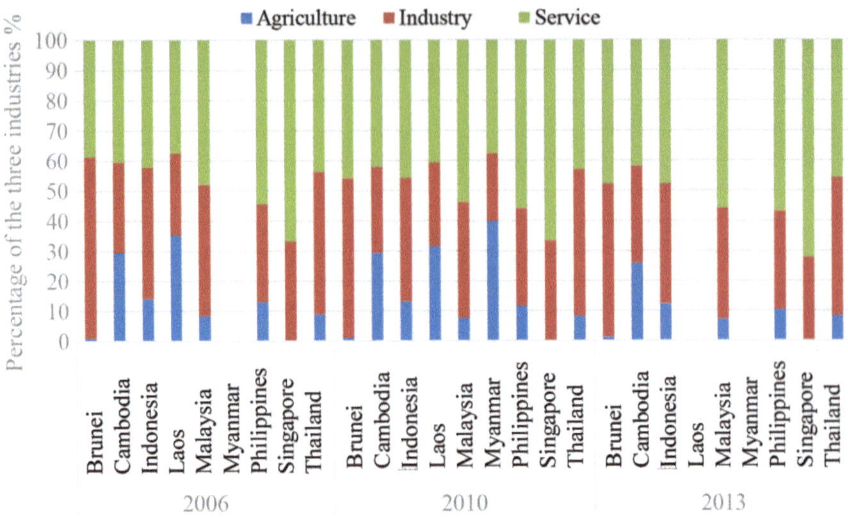


**Fig. 2.29** Industrial structure of ASEAN Member States. *Source* ASEAN Statistical Yearbook 2014

and national heritage park projects have been implemented in Laos, the Philippines, Myanmar, Thailand, Malaysia, and Vietnam, with the aim to protect land and obtain the 2010 biodiversity convention goals.

#### 2.2.3.2 Water Resources

According to Fourth World Water Development Report issued by the United Nations Educational, Scientific and Cultural Organization (UNESCO) in March 2012, 70% of the earth surface is covered by ocean, and freshwater resources are limited and unevenly distributed, with only 2.5% available for the mankind, animals, and plants. Located in the tropics, ASEAN boasts a great many rivers and lakes and rich freshwater resources.

Statistics indicates in 2007, ASEAN possessed 5674.5 billion $m^3$ of renewable water resources (see Table 2.6). Among the member states, Indonesia, Myanmar, and Malaysia rank top, with the total freshwater resources being 2838 billion $m^3$. However, Laos, Brunei, and Malaysia take the leading position for per capita freshwater resources, and Laos ranks top with 33,063 $m^3$/person·a. The demand for water resources rises along with the increase of population and the development of economy. ASEAN water consumption is expected to double in the second half of the twenty-first century.

Marine protected areas (MPAs) are established. In the past decades, MPAs have witnessed an increase in their quantity and area in ASEAN. According to UN statistics, ASEAN Member States possessed 87,778 $km^2$ of MPAs in 2007, 119.4%

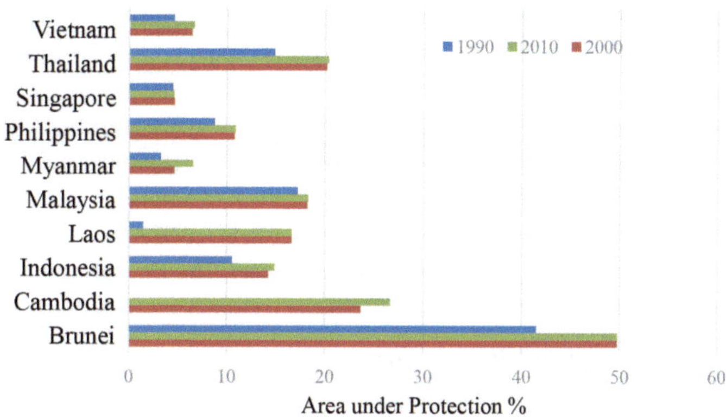


**Fig. 2.30** Ratios of land under protection. *Source* FAO Database, http://faostat.fao.org/site/684/default.asp ASEAN Statistical Yearbook 2014

higher than the 40,000 km$^2$ in 1995. Especially in 2001–2003, ASEAN's area of MPAs increased rapidly. Up to September 2009, the Philippines has established 339 MPAs, including 7000 islands, ranking first in terms of MPA quantity; Indonesia has 129 MPAs, ranking the second; the third was Malaysia, with 83 MPAs; Cambodia, Singapore, and Myanmar possessed few MPAs; and there was no MPA in Brunei, although two areas with coral reefs, Pelong Rocks and Pulau Punyit, are managed as wildlife sanctuaries (see Table 2.7). Seen from the area of a single MPA, the Savu Sea Marine National Park of Indonesia is the largest one in ASEAN, with an area of 35,000 km$^2$ [11].

**Table 2.6** Renewable freshwater resources of ASEAN Member States, 2007

| Country | Gross freshwater resources/Million m$^3$ | Freshwater resources per capita/[m$^3$/(person·a)] |
|---|---|---|
| Brunei | 85 | 22,254 |
| Cambodia | 1206 | 8493 |
| Indonesia | 28380 | 15,500 |
| Laos | 1904 | 33,063 |
| Malaysia | 5800 | 22,211 |
| Myanmar | 8806 | 18,202 |
| Philippines | 4790 | 5553 |
| Singapore | 9 | 194 |
| Thailand | 2100 | 3310 |
| Vietnam | 3665 | 4251 |
| ASEAN | 56745 | 57,385 |

*Source* Food and Agriculture Organization of the United Nations (FAO)

Table 2.7 ASEAN Member States' MPAs

| Country | Number of MPAs | Country | Number of MPAs |
|---|---|---|---|
| Brunei | 0 | Philippines | 339 |
| Cambodia | 2 | Singapore | 2 |
| Indonesia | 129 | Thailand | 23 |
| Malaysia | 83 | Vietnam | 36 |
| Myanmar | 6 | ASEAN | 620 |

*Source* Fourth ASEAN State of the Environment Report 2009

## 2.2.4 *Environmental Protection*

### 2.2.4.1 Atmospheric Environment

Air quality of some ASEAN Member States is worrying along with the use of biomass for energy, the increase of population, and the quickening of urbanization. In particular, transport vehicles, similar with industry, have been among the main sources of air pollution in ASEAN Member States.

In Indonesia, air quality was "unhealthy" for 49 days in Djakarta, 18 days in Medan, and 7 days in Surabaya in 2007. In other densely populated cities, the increase of motor vehicles has brought up the concentration of nitrogen compounds up 30 ppm in average. Moreover, another main reason for the air quality degradation in Indonesia is the development of industry, especially the sectors of food, chemicals, petroleum, coal, rubber, plastics, papermaking, and textile.

In Brunei, $PM_{10}$ concentration grew from 12.3 $\mu g/m^3$ in 2006 to 18.1 $\mu g/m^3$ in 2008, though its air quality was defined "good" for every day of 2008.[6]

In Malaysia, air pollutants include mainly nitrides, sulfides, and suspended substances (PM). Since the end of the 1990s, Malaysia has achieved a significant improvement in air quality, with average $SO_2$ concentration dropping from 0.0074 ppm in 1998 to 0.0019 ppm in 2007, and $PM_{10}$ basically consistent with state standard between 2000 and 2008. Traffic and electric power sectors are major sources of air pollutants, with nitrides from the former and sulfides and suspended substances from the latter. According to statistics, the electric power sector contributes 60% of $SO_2$ and 50% of PM, and the traffic sector contributes most nitrides and 35% of PM.

In Thailand, urban air pollution is largely owing to transport vehicles, and 50% of the fuel is consumed in Bangkok. Statistics indicates in 2007, the overall air quality of Thailand was "good" for 141 days, "medium" for 197 days, and "unhealthy" for 27 days; moreover, most pollutants in Bangkok, the capital, were particulate matters, especially inhalable ones, and the content of CO, $NO_x$, $SO_2$, HC, lead, and black smoke exceeded the state standard. According to statistics, in the pollutants of Bangkok, 54% of hydrocarbons and 88.2% of carbon compounds

[6]The state standard is 50 $\mu g/m^3$ in Brunei.

are discharged from motorcycles. Total particulate matter concentration of Bangkok is far higher than the state standard, 50 ppm, though it has improved since 1995.

In the Philippines, air pollution is attributed largely to the emissions of transport vehicles. According to the prediction, 65% of the pollutants are derived from mobile sources, 21% from fixed sources, and the other 14% from regional sources. Despite the drop, TSP concentration remains higher than the state standard 90 ppm.

Singapore registers good air quality. In 2008, the air quality was "medium" for 353 days, and other indexes, except $PM_{2.5}$, were consistent with the UN standard. In Vietnam, air pollution is caused largely by traffic and industrial sectors. As the largest source, traffic contributes about 70% of urban air pollution. Across the country, transport vehicles discharge 85% of the CO and 95% of the VOCs.

**Air quality management**

ASEAN and ASEAN Member States have formulated relevant laws, promulgated pertinent policies, and implemented various measures, with the aim to control air pollution and improve air quality. To prevent the further worsening of air quality, Malaysia, Singapore, and Thailand introduced measures for restricting the use of lead-containing petrol as early as 1991, and Thailand completed the objective and stated to stop the use of lead-containing petrol in 1996. Indonesia started to bring in the measure in 2001 and completed the task in 2006. At present, all ASEAN Member States have stopped using lead-containing petrol, except Cambodia, Laos, and Myanmar.

ASEAN Member States adopt differentiated policy measures to control air pollution and improve air quality. For instance, Cambodia lays particular emphasis on clean energy and energy efficiency; Laos focuses on the mitigation of the excessive reliance on traditional energy and then the improvement of utilization rate of renewable energy; Thailand stresses the reduction of pollutants emitted from transport vehicles and takes the improvement of renewable energy utilization rate through the development of ethanol industry as a national strategy; and Singapore controls air quality through conducting monitoring strictly, formulating land use plans prudently, separating industrial zones from residential areas, and introducing European transport vehicle emissions standard compulsorily and etc.

To monitor air quality better, most ASEAN Member States have established a monitoring network to get the air quality status in real time. Singapore has set up the first air quality monitoring station (in 1971), followed by Malaysia (in 1978). Myanmar also built up the network in 2008. Up to 2009, there had been 177 air quality monitoring stations in ASEAN Member States for monitoring the atmospheric pollutants, such as PM, CO, $NO_x$, $SO_x$, Pb, $O_3$, and TSP.

ASEAN Agreement on Transboundary Haze Pollution (ASEAN 2002) aims to monitor and prevent the haze pollution caused by land and/ or forest fire and approves the adoption of the zero-burning policy. ASEAN Agreement on Transboundary Haze Pollution is among the international agreements intended to avoid the international communication of particulate matters generated in forest combustion [12].

#### 2.2.4.2 Water Environment

ASEAN Member States are bothered with both water pollution and the increasing demand for water (see Table 2.8).

In 2007, 27 of the over 30 rivers monitored in Indonesia were contaminated; and in 2008, 54% of the rivers monitored were polluted. In Thailand, rivers also suffer heavy pollution. In 2005, 29% of the rivers there witnessed "poor" water quality and the rate rose to 48% in 2008. In Malaysia, mildly contaminated rivers are largely located where agricultural and industrial production activities concentrate, and most contaminated rivers are situated in ports and industrial production-concentrated areas. Few rivers are polluted where forest coverage is high and industrial development does not pick up the pace. In Philippines, water quality of all the rivers monitored is unoptimistic and many waters are under the state standard regarding COD and BOD. Water pollution is serious as well in Vietnam.

According to the ASEAN Statistical Yearbook 2014, ASEAN Member States differ a lot from one another in terms of the ratio of safe drinking water-available population (see Fig. 2.31), which is 100% for Brunei, Singapore, and Thailand, 94% for Malaysia, 91% for Vietnam, 83% for the Philippines and Myanmar, 70% for Laos, 54% for Cambodia, and 41% for Indonesia.

**Table 2.8** Water quality in selected ASEAN Member States

| Country | Year | Water quality |
|---|---|---|
| Indonesia | 2008 | 54% of 33 rivers monitored were heavily polluted |
| Philippines | 2008 | 14–28% of the rivers exceeded the BOD emission limit |
| Thailand | 2008 | Quality of 48% of the rivers was poor in 2007 and 29% in 2005 |
| Vietnam | 1996–2001 | BOD emission of the rivers was 2–3.8 times the state standard |

BOD = Biological Oxygen Demand

*Note* BOD is the amount of dissolved oxygen needed for the aerobic biological tissues in a water to resolve the organic substances in a water sample in a certain term and at a specific temperature. Usually, BOD value is used to indicate the organic contamination degree of waters

*Source* [a]ASEAN Secretariat, 2009. The Fourth ASEAN State of the Environment Report. http://www.aseansec.org/publications/SoER4-Report.pdf. [b]Asian Development Bank, 2007. 2007 Asian Water Development Outlook. Manila. http://www.adb.org/publications/asian-water-development-outlook-2007; [c]World Health Organization (WHO) and United Nations International Children's Emergency Fund. 2010. Health and Drinking Water Development Report. Geneva: World Health Organization. http://www.unicef.org/eapro/JMP-2010Final.pdf (Access Date, November, 2013)

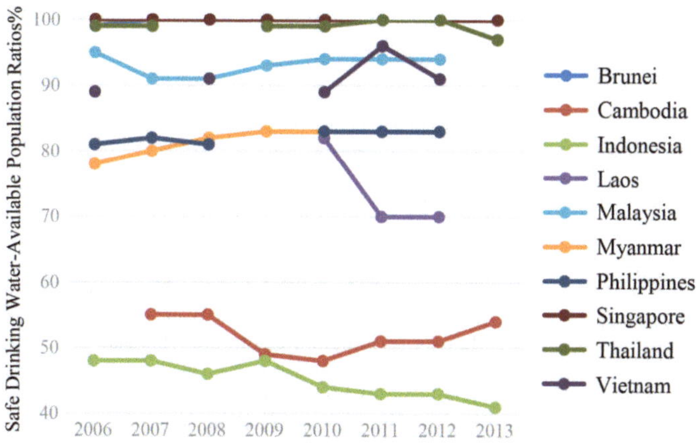


**Fig. 2.31** Safe drinking water-available population ratios of ASEAN Member States, 2006–2013. *Source* Data that AMS submitted for ASEAN Statistical Yearbook and ACPMS Report; Cambodia—Cambodia Social and Economic Survey and Cambodia Demographic and Health Survey; Malaysia—Water Works Department, Rural Environment Sanitation Project (RESP), Population and Housing Census; The Philippines—National Demographic and Health Survey; Singapore—Public Utilities Board and administrative records; Thailand—Population and Housing Census; Vietnam—Living Standard Survey, Multiple Indicator Cluster Survey and Multi-purpose Household Survey. Key indexes of Asian Development Bank for 2006–2014; UN Statistics Division-population and social database, 2009 Asian-Pacific Economic and Social Yearbook. *Note* '—' No information was available at the time of publication. ASEAN Statistical Yearbook 2014

#### 2.2.4.3 Solid Wastes

**Municipal solid wastes**

Municipal solid wastes (MSWs) are generally defined as household garbage and harmless wastes, such as commercial and institutional wastes, street sweepings and building rubbish, human feces, dust from incinerators, and sludge of digestion tanks and sewage treatment works.

MSWs are a major part in the wastes of most ASEAN Member States (see Table 2.9). MSWs account for approximately 67% of the gross wastes in Thailand. In Bangkok, the metropolitan area and surrounding provinces generate 30% of its total MSWs. Favorable progress has been achieved in the reduction of wastes through garbage recycling projects and the provision of safe and efficient garbage collection and treatment systems. From 2005 to 2007, MSWs in Thailand decreased by 13% every day on average.

Annual output of wastes in urban areas of Myanmar remained basically unchanged in the past four years, steadily at 700,000 tons. However, daily generation of wastes of Rangoon tripled, from 564 tons in 1990 to 1324 tons in 2007. Over the past few years, Singapore formulated various strategies for reducing and recycling wastes, with the aim to bring down the quantity of wastes buried. The

**Table 2.9** Municipal solid wastes generation in selected ASEAN Member States, 2005–2008

| Country | 2005 | 2006 | 2007 | 2008 |
|---|---|---|---|---|
| Malaysia[a] | 32.90 | 26.50 | 11.40 | 11.40 |
| Burma[b] | 0.69* | 0.70* | 0.69* | 0.71* |
| Philippines[c] | – | – | 12.15 | – |
| Singapore[d] | 5.01^ | 5.22^ | 5.60^ | 5.97^ |
| Thailand[e] | 7.64 | 6.82 | 6.64 | |

*Source* [a]Ministry of Environmental Protection of Indonesia
[b]Ministry of Forestry of Myanmar
[c]Philippines National Solid Waste Management Committee
[d]Ministry of Environment and Water Resources of Singapore
[e]Ministry of Pollution Control of Thailand
*Note* *Data of Rangoon and 238 towns (exc. Mandalay and Naypyidaw) of Myanmar
^Including the quantity of wastes recycled, incinerated and buried

capital of Philippines generates the most solid wastes (24% or 2.86 tons). According to the prediction, MSWs in Philippines would amount to 13.67 million tons by 2010. In 2005–2008, Indonesia witnessed a remarkable reduction in MSWs. The MSW management regulations Indonesia has issued recently are helpful in promoting 3Rs and the treatment and utilization of wastes, identifying the role of communities, introducing incentive and inhibition mechanisms, and clarifying authority and division of work.

In the ASEAN region, WSMs mainly include organic wastes, plastics, paper and paperboards, textiles, rubber and leather, timber, glass, and metal (see Table 2.10). On average, organic wastes account for a proportion of 46%, followed by plastics (18%) and paper (14%). The composition is expected to change, along with the rise of urbanization rate and income level.

Most MSWs are buried in a sanitary manner or stacked in the open air (see Fig. 2.32). However, manure mixture for fertilizing, incineration, and other methods are increasingly applied, including material recycling facilities (MRFs) and refuse-derived fuel plants (RDFPs). In Malaysia, there are 10 sanitary landfills and 188 open garbage dumps as well as 5 waste recycling centers and 1 refuse-derived fuel facility. Up to 2007, there had been 700–800 open garbage dumps in Philippines. Many local governments have determined to replace open dumps with sanitary landfills. In 2007, there were 2200 MRFs serving 2473 towns. In Singapore, about 56% of the solid wastes (including industrial wastes) are recycled and 41% are burned in wastes-based power plants. The other 3%, incombustible wastes, are delivered to Pulau Semakau Offshore Refuse Landfill for treatment. According to the policies, all combustible wastes must be burned in Singapore, which can reduce the quantity of wastes by 90%. Meanwhile, the heat from the burning is used to generate electricity, accounting for 2% of the electricity supply in the country.

**Table 2.10** MSW components in selected ASEAN Member States

| Country | Components/% | | | | | | | | |
|---|---|---|---|---|---|---|---|---|---|
| | Organics | Plastics | Paper/Paperboards | Textiles | Rubber and Leather | Timber | Glass | Metal | Other |
| Brunei[a] | 42 | 16 | 18 | 0 | 0 | 0 | 3 | 4 | 17 |
| Indonesia[b] | 58 | 14 | 9 | 2 | 2 | 2 | 2 | 2 | 6 |
| Laos[c] | 30 | 30 | 15 | 0 | 0 | 0 | 25 | 0 | |
| Malaysia[d] | 49 | 17 | 10 | 0 | 0 | 0 | 4 | 2 | 0 |
| Burma[e] | 73 | 2 | 18 | 2 | 0 | 4 | 0 | 0 | 1 |
| Philippines[f] | 50 | 25 | 13 | 0 | 0 | | 3 | 5 | 5 |
| Singapore[g] | 24 | 24 | 25 | 3 | 3 | 2 | 2 | 17 | |
| Thailand[h] | 42 | 14 | 16 | 3 | 1 | 7 | 5 | 3 | 9 |
| Vietnam[i] | 49 | 16 | 2 | 1 | 7 | 6 | 19 | | |

*Source* [a]Ministry of Environment, Parks and Recreation and Ministry of Development, Brunei
[b]Ministry of Environment, Indonesia
[c]Environmental monitoring of Laos in 2005
[d]Ministry of Solid Wastes Management, Malaysia
[e]Ministry of Forestry, Myanmar
[f]Environmental Administration Bureau and Ministry of Environment and Natural Resources, Philippines
[g]Ministry of Environment and Water Resources, Singapore
[h]Bureau of Natural Resources and Environmental Policies and Planning, Thailand
[i]Ministry of Natural Resources and Environment, Vietnam

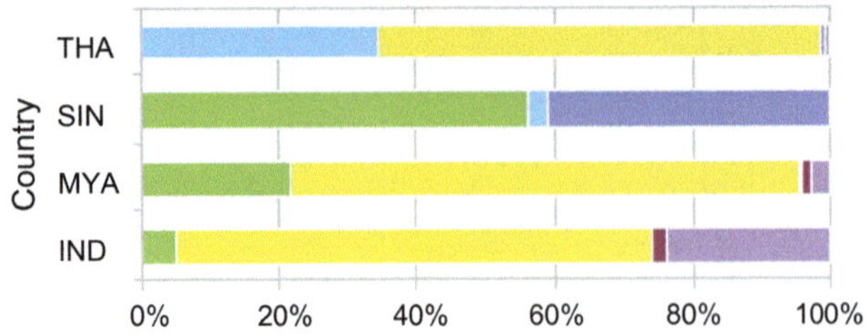


| | IND | MYA | SIN | THA |
|---|---|---|---|---|
| Recycling | 5 | 22 | 56 | 0 |
| Sanitary landfill | 0 | 0 | 3 | 34.5 |
| Open dumps | 69 | 73 | 0 | 64 |
| Incineration | 0 | 0.7 | 41 | 1 |
| Composting | 2 | 1.3 | 0 | 0 |
| Others | 24 | 3 | 0 | 0.5 |

**Fig. 2.32** MSW treatment and disposal methods in selected ASEAN Member States. *Source* Ministry of Environment, Indonesia; Ministry of Forestry, Myanmar; Ministry of Environment and Water Resources, Singapore; Natural Resources and Environment Policies and Planning Bureau, Thailand. *Note* Data of Singapore are about all types of wastes. Recycling methods include manure mixture for fertilizing and material recycling facilities

**Industrial solid wastes**

The largest source of industrial wastes is the manufacturing of basic metals, tobacco, timber and woodware, and paper and paper products in Thailand, Singapore, and Malaysia (see Fig. 2.33). In 2000, ASEAN is projected to produce 19 million tons of industrial wastes.

Between 2002 and 2008, Thailand generated 1.45 million tons of industrial wastes every year on average. Singapore treated merely 1 million tons of industrial wastes annually, though its total generation of industrial wastes was 3 million–4.5 million tons, mainly due to the implementation of many recycling measures in the country. In 2008, Malaysia produced 1.3 million tons of industrial wastes.

Hazardous wastes account for 1–3% of the wastes generated in ASEAN Member States. It's estimated that hazardous wastes amounted to 3 million tons in 2000, which is much higher now, because manufacturing and agricultural sectors generated nearly 4 million tons of hazardous wastes in 2007. Besides, domestic and commercial activities will also generate a handful of hazardous wastes.

In some ASEAN Member States, industrial wastes are still collected, transported, treated, and disposed as MSWs. Finally, they get mixed with household wastes and discarded at open dumps and landfills. There is still illegal dumping in some states due to insufficient facilities and high cost of treatment. However, some ASEAN Member States have established industrial waste treatment facilities. For instance, about 51% of industrial wastes are treated locally before disposal in Malaysia. Other main methods include refuse-derived fuel plants, sanitary landfills, and incineration.

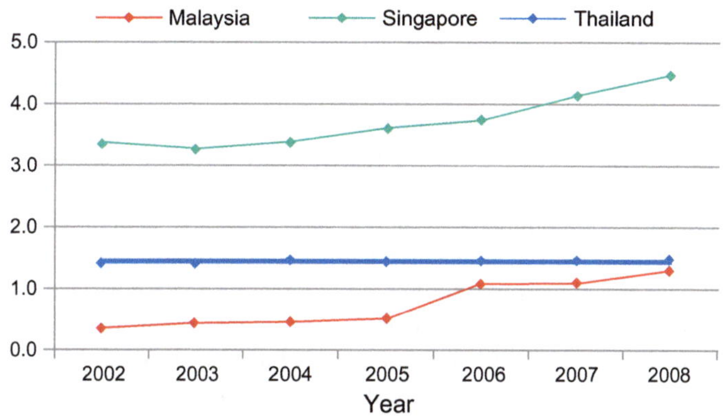


**Fig. 2.33** Generation of industrial wastes in Thailand, Singapore and Malaysia. *Source* Ministry of Environment, Malaysia; Ministry of Environment and Water Resources, Singapore; and Ministry of Pollution Prevention and Control, Thailand

## *2.2.5 Energy and Climate Change*

### 2.2.5.1 Energy Consumption

According to the BP Statistical Review of World Energy 2012, among the petroleum reserves detected in ASEAN Member States, Malaysia possesses the largest reserve, which is 5.9 billion barrels, followed by Vietnam, Indonesia, Brunei, and Thailand in succession. Up to the end of 2011, natural gas reserves were concentrated in Indonesia, Malaysia, Vietnam, Brunei, Thailand, and Myanmar. Coal resources are located mainly in Indonesia, Thailand, and Vietnam. Indonesia boasts of producing 5.529 billion tons of coal, 0.6% of the global reserve. In addition, there are also coal reserves in Malaysia, Laos, and Philippines. In addition to coal reserve, the ASEAN region possesses abundant hydropower resources and wood fuel resources.

On the whole, fossil fuel continues to dominate the energy structure of ASEAN. Gross energy use keeps growing along with the constant increase of population, and GDP and its growth stays consistent with the increase of per capita income and the middle class emerging in the region. From 1990 to 2007, gross energy use of ASEAN doubled, from 253 million tons of oil equivalent to 511 million tons of oil equivalent. In the meanwhile, the use of coal and natural gas increased and the fossil fuel-to-gross energy demand rate rose from 55.7 to 72.4% while the ratio of petroleum remained basically unchanged. The use of non-fossil fuel witnessed a decrease because commercial fuel substituted traditional energy and the proportion of biomass fuel (generally including forest and agricultural residue) dropped from 40.7 to 23.5%.

The growth of economy and the improvement of living standards have increased the demand for electricity. Therefore, investment is needed to develop electrification in rural areas and the most undeveloped states and provinces of ASEAN. In 2010, the electrification rate exceeded 90% for the rural areas of ASEAN and 55% for rural areas. However, the rate dropped to 66% for urban areas and merely 12.5% for rural areas in Cambodia. Besides, power grid should be expanded so as to provide sufficient, reliable, and properly managed electric power system consistent with the expected development scale.

The expansion of commercial energy has promoted economic growth and poverty reduction. However, it has also intensified constantly the reliance on fossil fuel, aggravated global warming further, and increased ASEAN risks of exposure to world petroleum prices. In the past decades, petroleum prices experienced several increases, which not only burdened consumers but also delayed the development. A reason lies in the surging demand from Asian countries, though the demand from developed markets remains weak. In the future, it is unlikely that petroleum prices tend to stabilize. Brunei and Malaysia, who are net oil exporters, may get benefited from the high prices, and most Southeast Asian countries will be on the contrary. On the whole, import outgrows export. In 2009, the net import volume of oil was 42.5% of the gross consumption. To make things worse, several governments have increased the subsidies for the retail prices of oil, incurring a financial burden and reducing the energy conservation pressure of enterprises and households. For long-term development, the fully market-oriented fuel pricing mechanism should be adopted, though policy makers have decreased relevant subsidies.

ASEAN primary energy structure see Fig. 2.34.

The prices of petroleum and other fossil fuels staying high give an impetus to the development of renewable energy, especially biofuel, geothermal energy, hydropower, solar energy, and wind energy. ASEAN Member States are exploring proactively the use of renewable energy or other alternative energy and have put forward the renewable energy development plan. For instance, Indonesia is projected to raise the renewable energy-to-national energy structure rate from 4.79% in 2011 to 25% in 2025; the Philippines hopes to be able to satisfy half the demand for electricity with renewable energy by 2030; the renewable energy goal of Malaysia is to raise the renewable energy-to-energy mix rate to 11% by 2020; Vietnam plans to increase the proportion of new energy and renewable energy to about 5% by 2020 and to 11% by 2050. ASEAN's switch from the reliance on fossil fuel and the traditional "black economy" will impose great influence on the oil-depending countries. The local production and supply of renewable energy, such as biofuel, can help to mitigate the reliance on crude oil. However, under the influence of such factors as technological restriction, financing constraint, and insufficient financial arrangement, ASEAN Member States are confronted with various barriers for the development of renewable energy so that their actual action lags always behind the objectives [13].

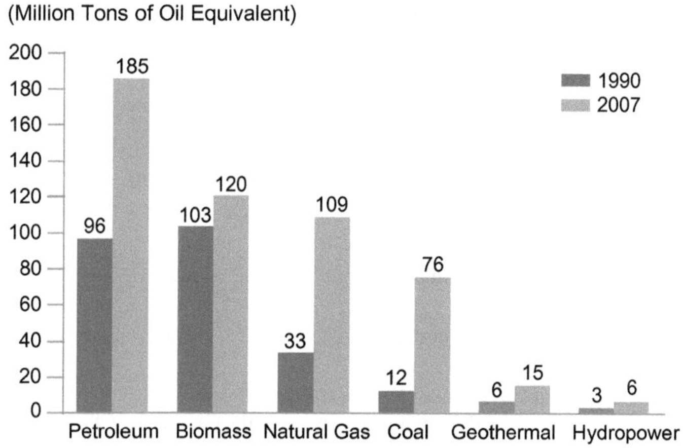


**Fig. 2.34** ASEAN primary energy structure. ASEAN = Association of Southeast Asian Nations. *Note* Ton of oil equipment represents the amount of energy released in the burning of 1 ton of crude oil. *Source* Chira Achayuthukan and Weerakom Ongsakul. 2012. ASEAN 2030 Energy Demand. Background documents for the ASEAN 2030 research and formulation

#### 2.2.5.2 Climate Change

According to The Economics of Climate Change in Southeast Asia: A Regional Review by Asian Development Bank, Southeast Asia is the most vulnerable to climate change. The report indicates that in the past 50 years, the average temperature of Southeast Asia has risen by 0.1–0.3 °C every 10 years and its sea level increases by 1–3 mm every year. Compared with other regions, the ASEAN region witnesses a small increase and a slow growth in its annual average emission of greenhouse gases (for a major reason of global warming) (see Table 2.11). For instance, the annual average increment of $N_2O$ and $CH_4$ was lower in 2010 than the years before.

### 2.2.6 Ecosystem and Biodiversity

#### 2.2.6.1 Change in Forest

Located in the tropics, the ASEAN region is among the regions with the most dense and most extensive forest coverage. The abundant forest resources bring forth the ample reserves of timber, biodiversity, and carbon resources and also play an important role in improving the ecology, environment, and climate of the region. In the ASEAN region, forests cover an area of nearly 213 million ha, for which Indonesia, Myanmar, and Malaysia rank top (Table 2.12) [14].

**Table 2.11** ASEAN greenhouse gas emissions

| Emission of atmospheric pollutants | | 1995 | 2000 | 2005 | 2010 |
|---|---|---|---|---|---|
| $N_2O$ | $CO_2$ equivalent/1000 metric ton | 16,313 | 18,029 | 20,229 | 20,633 |
| | Annual average growth after the previous period/% | | 2.1 | 2.4 | 2.0 |
| $CH_4$ | $CO_2$ equivalent/1000 metric ton | 48,135 | 517,670 | 527,670 | 53,030 |
| | Annual average growth after the previous period/% | | 1.5 | 0.4 | 0.1 |
| CFCs | Ozone consumption potential/metric ton | 21,944 | 14,318 | 1037 | 1025 |
| | Annual average growth after the previous period/% | | −7.0 | −18.6 | −0.2 |

$CH_4$ = methane; $CO_2$ = carbon dioxide; $N_2O$ = nitrogen dioxide
*Source* Intergovernmental Panel on Climate Change (IPCC), 2007. Climate Change 2007: Synthesis Report. Geneva. http://www.ipcc.ch/pdf/assessment-report/ar4/syr/ar4_syr.pdf (Access Date: October, 2013)

**Table 2.12** ASEAN forest area and woodland area

| Country | Forest/1000 $hm^2$ | Other woodland/1000 $hm^2$ | Total/1000 $hm^2$ | Coverage rate/% |
|---|---|---|---|---|
| Brunei | 380 | | | |
| Singapore | 2 | | | |
| Cambodia | 10,094 | 133 | 10,227 | 57.9 |
| Indonesia | 94,432 | 21,003 | 115,435 | 63.7 |
| Laos | 15,751 | 4834 | 20,585 | 89.2 |
| Malaysia | 20,456 | 0 | 20,456 | 62.3 |
| Myanmar | 31,773 | 20,113 | 51,886 | 79.4 |
| The Philippines | 7665 | 10,128 | 17,793 | 59.7 |
| Thailand | 18,972 | 0 | 18,972 | 37.1 |
| Vietnam | 13,797 | 1124 | 14,921 | 48.1 |
| ASEAN | 213,322 | 57,385 | 271,449 | 62.7 |

*Note* Coverage rate denotes the proportion in the national gross of woodland area
*Source* Food and Agriculture Organization of the United Nations (FAO), 2010. World Forest Resources Assessment

**Forest area decrease**

Forest is the most abundant resource of ASEAN, but is also faced with similar challenges (see Fig. 2.35). The increasing demand for timber, fuel wood, and other forest products as a result of fast population growth and the conversion of forest land to agricultural lands are damaging the forests in this region. The rapid destruction of forests in Southeast Asia reduces the region's resilience to and exposes it to the impacts of global warming. Between 1990 and 2010, the forest area of ASEAN decreased by 2%. As for Cambodia, the forest coverage rate has

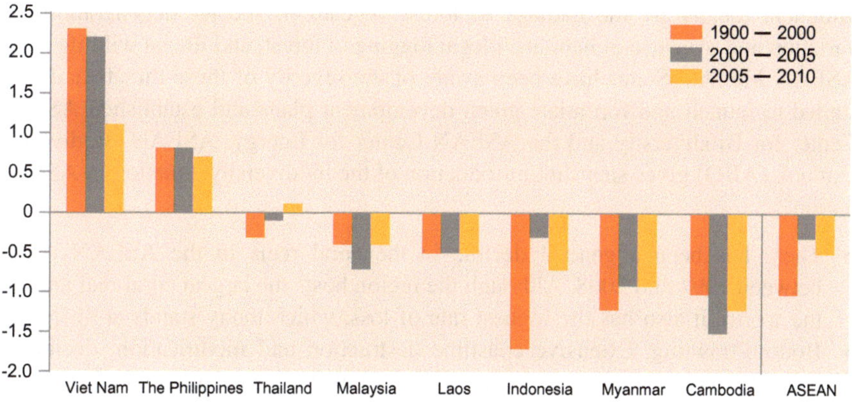


**Fig. 2.35** ASEAN forest area variation, 1990–2010. ASEAN = Association of Southeast Asian Nations. *Source* Food and Agriculture Organization of the United Nations (FAO), 2010. World Forest Resources Assessment. Rome. http://www.fao.org/docrep/013/i1757e/i1757e.pdf (Access Date: November, 2013)

fallen to 52%, from 73% in 1965. The drop results in the decrease of species and indirectly causes the reduction of contribution to $CO_2$ emission reduction [15].

The decrease in ASEAN forest area is largely attributable to the growing population, the increase of agricultural production, logging and mining, for the foreign exchange earning of many ASEAN Member States depends still heavily on the export of timber and agricultural products. Furthermore, ASEAN Member States have been always threatened with illegal logging owing to the lack of adequate resources allocated to monitoring and law enforcement. They witness decreases in the area of mangrove, except Brunei, basically unchanged. ASEAN region has very rich mangrove resources, but its decrease rate tops the world. In the past several decades, mangroves are decreasing by 628 $km^2$ each year from 63,850 $km^2$ in total in 1980 to 46,971 in 2005, with a total reduction of 26% in 25 years.[7]

#### 2.2.6.2 Biodiversity

ASEAN accounts for a measly 3% of the global land area but is among the regions with the most abundant biological resources. It hosts close to 18% of the species assessed by the IUCN. 3 of the top 17 countries with rich biodiversity are situated in ASEAN (Indonesia, Malaysia, and the Philippines), with above 70% of world biodiversity. However, such severe problems as the damage to natural resources and the degeneration of species and their habitats are confronting ASEAN. Preliminary calculation indicates more than 1000 species have been put in

[7]China-ASEAN Environmental Cooperation Center, Implementation Progress of China-ASEAN Biodiversity Strategy and Action Plan, China Environmental Science Press, 2016.

imminent danger by the damage to forest vegetation, species degeneration, the variation and over-use of habitats, illegal logging of forest, and illegal wildlife trade. ASEAN Member States have been aware of the severity of these threats and have started to launch and formulate green development plans and established ASEAN Center for Biodiversity and the ASEAN Center for Energy. ASEAN Biodiversity Outlook (ABO) gives snapshot introduction of the biodiversity situation in ASEAN region[8]:

- There has been a general decline in the coral reefs in the ASEAN region between 1994 and 2008. Although the region hosts the largest coral reef areas in the world, it also has the highest rate of loss, which today stands at 40%.
- Bottom-trawling, extensive coastline destruction and modification, decline in coastal water quality, and human-induced development have endangered seagrass beds in the ASEAN region. Indonesia, the Philippines, Singapore, and Thailand have each experienced from 30 up to 50% losses of seagrass habitats, compounded by the fact that the loss figures for other Southeast Asian countries remain largely unknown.

#### 2.2.6.3 Ecological Footprint

Ecological footprint (EF) is an index for measuring human demand for global ecosystem services and nature's capacity to meet these demands. According to the analysis done by Global Footprint Network (GFN) in 2005, ASEAN region's EF is merely 4% of the global footprint, which indicates that ASEAN's share of global resource consumption is lower than its share of global population (9%) (see Fig. 2.36). However, ecological deficits occur in 5 ASEAN Member States, while the number was only 3 when the 3rd ASEAN State of the Environment Report was issued. Measures should be taken to make a changeover to ensure a more sustainable future of the ASEAN region.

Evidently, ASEAN economy needs a transition in the modes of production and consumption. ASEAN is capable of initiating actions for realizing green economic transition. The action includes formulating further sustainable measures and popularizing environmental products and service. Along with the continuous development of ASEAN, various measures are expected to stimulate the development of green cities, the use of clean and efficient techniques and the application of renewable energy and thereby ensure the integrity of the ecosystems.

---

[8]China-ASEAN Environmental Cooperation Center, Implementation Progress of China-ASEAN Biodiversity Strategy and Action Plan, China Environmental Science Press, 2016.

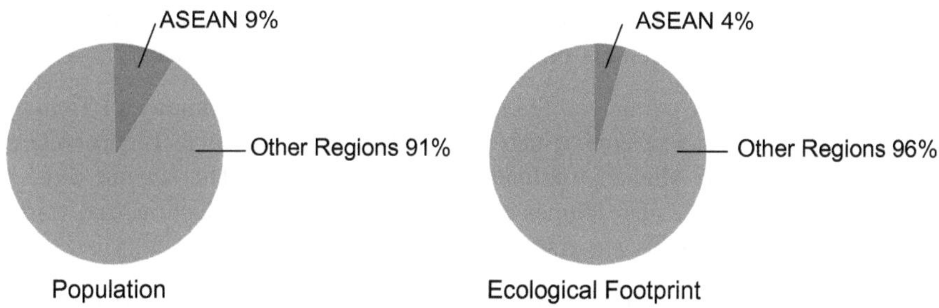


**Fig. 2.36** ASEAN population and ecological footprint. *Source* Global Footprint Network (GFN)

## *2.2.7 Sustainable Production and Consumption*

### 2.2.7.1 Eco-Label

The eco-label plan has been implemented in many ASEAN Member States to encourage sustainable production. On June 5, 2004, Indonesia initiated the eco-label certification and accreditation scheme on the World Environment Day, with the certification authority approved by the national accreditation service committee and the standard based on ISO 14024, other eco-label plans, legal requirements, relevant international conventions and yield and quality standards.

In 2009, Vietnam approved its national eco-label plan, with the objectives of:

- Protecting and efficiently using available natural resources;
- Encouraging environmental improvement; and
- Encouraging clients to make wiser decisions.

It plans to make 100% of imported products and 50% of domestically made products printed with the ISO 14021 environmental label.

Singapore Green Label Scheme (SGLS) was initiated by the Ministry of Environment in May 1992 and is applicable to most products, excluding foods, drinks, and pharmaceuticals. Green labels are applicable to the products consistent with the ecological standard as stated in SGLS or mutually recognized and supported by SGLS, for which the authentication by other member institutions is allowed. Singapore Environment Council (SEC) has implemented it since June 1999, and a new SCLS standard has been established according to the markets of Singapore and foreign countries and consumers' demand for products. Green label access standard is established by:

- The advisory committee made up of representatives of governments, private sector, academic institutions, and statutory organs;
- The uniform standards formulated by the committee based on the industrial investigation conducted by manufacturing enterprises that give an active response; and

- The opinions given by GEN20 and its member institutions on standard formulation and the agreement on the mutual recognition of eco-labels.

In October 1993, the Commercial Sustainable Development Committee of Thailand initiated the green label scheme, which was launched officially by its Environmental Research Institute of Ministry of Industry in August 1994. The scheme aims to generalize the concepts of resources conservation, pollution reduction, and wastes management. Green labels are granted for the purpose of:

- Providing reliable information to guide customers how to select products;
- Providing consumers with the opportunity of making environmental decisions and developing and providing more environment-friendly products and establishing a market incentive mechanism for production enterprises; and
- Reducing the environmental influence that may arise from the production, utilization, consumption, and disposal of products.

The green label scheme of Thailand is applicable to products and services, excluding foods, drinks, and pharmaceuticals. Any product or service that meets the green label standards will be labelled accordingly. Participation in the scheme is on a voluntary basis.

#### 2.2.7.2 Sustainable Production

Sustainable production will be helpful in reducing the use of resources and energy in the production of products and services while minimizing the impact on the environment. In addition, it will make products safe, ecologically harmless, durable, repairable, recyclable, fertilizing, and degradable.

In the meantime, the entire life cycle—from the extraction of natural resources to the production, distribution, and disposal of products and services—is involved. The sustainable production in key economic sectors, i.e., agriculture, manufacturing, forestry, and energy, will help conserve sufficient resources for future generations.

**Sustainable agriculture**

The intensification of agricultural production has given rise to various environmental problems, such as soil erosion and degradation, the increase of greenhouse gas emission, biodiversity loss, and pollution aggravation. It is urgently necessary to adopt sustainable modes of agricultural production so as to maintain environmental health and promote social and economic justice while guaranteeing economic profit. As agriculture serves as the foundation of many ASEAN economies, society, enterprises, and environment will benefit substantially from the participation in such agricultural certification projects as Roundtable on Sustainable Palm Oil (RSPO) and Rainforest Alliance.

Over three decades, oil palm industry has grown into a major engine for economy, especially in Malaysia and Indonesia. In recent years, the sustainable production of palm oil has speeded up as a mass of planting enterprises participate in RSPO certification. RSPO aims to promote the sustainable production and use of oil palm products and the participation of stakeholders in the supply chain through reliable global standards (embodied in its principles and standards). The emphasis of RSPO principles and standards is attached to environmental protection and also the well-being of local communities and planting workers. The standards include protecting forests and peatland with a high value, supporting the land right of local communities, improving health and the right to education, etc.

Producers of crops, besides oil palm, can also apply for such certificates as Reforest Alliance certification. To obtain the certification of the Rainforest Alliance, agricultural producers should abide by the best social and environmental practices as specified in sustainable agricultural network (SAN) standard. Once passing the certification, they can apply the Rainforest Alliance certification label to their products. The label indicates the producers' reduced ecological footprint and less adverse impacts on human beings and wildlife.

**Manufacturing—cleaner production**

Manufacturing is an important sector consuming resources and producing wastes in ASEAN region. However, it can be transformed into a driving force for sustainability by cleaner production and design of products and services with a better environmental performance. Heavily stressed by the strict environmental laws and regulations, manufacturing makes active efforts to reduce emissions and pollutants by adopting various control and treatment measures. Recently, actions improving environmental performance have shifted gradually to the transformation of lifestyles and the implementation of comprehensive environmental strategies and management regulations. Manufacturing enterprises also start to perform bigger environmental responsibilities in its whole value chain. The adoption of the comprehensive and systematic methods for improving sustainable performance lays a foundation for new operating modes or supply modes, and the latter are expected to bring forth remarkable environmental benefits.

**Sustainable forestry**

Forestry is an important economic activity of many ASEAN Member States, especially those rich in forests and commercial timber. Traditionally speaking, forest will provide people with clean water resources, food and medical materials as well as important social and cultural relations, especially for the indigenous people. In addition, forests play an important role in improving air quality and mitigating climate change. With proper management, forest and tree planting industry will benefit forestry-based populations and the overall global community. Therefore, it is necessary to make forestry activities (from logging to final products) sustainable [16].

The Forest Stewardship Council (FSC) has established the standard for sustainable forest production. It resorts to certification to promote local participation in forestry market, improve the awareness of forest value and thereby improve social and environmental standards in global forest management. ASEAN is striving to improve forest management and timber harvesting by reducing the adverse impact of illegal logging and recognizing forests' social, economic, and environmental values.

The enterprises intending to acquire FSC certification must adjust their management and operation so as to meet the social and environmental standards of FSC. FSC establishes 10 best practices to satisfy the social, economic, ecological, cultural, and spiritual demands of the present and future generations.

FSC follows the following principles in forest management:

- Compliance with laws and FSC principles;
- Tenure and use rights and responsibilities;
- Indigenous people's rights;
- Community relations and workers' rights;
- Benefits from the forest;
- Environmental impact;
- Management plan;
- Monitoring and assessment;
- Maintenance of high conservation value forests;
- Plantations (Accountable plantation management).

FSC audits each holder of its certificate once a year at least. ASEAN Member States have acquired FSC certification for 1,323,781 ha of forest in total (see Table 2.13), for which Indonesia ranks top, followed by Malaysia.

**Reduce, reuse, and recycle (3R)**

The 3R method is becoming popular. Most ASEAN Member States have launched activities to improve national awareness so as to win the recognition of the general public and encourage them to take part in 3R activities.

In Indonesia, many government institutions are committed to carrying out such activities as the reduction, reuse, and management capacity improvement of wastes. The management methods including manure mixture for fertilizing, recycling, and the redevelopment of recyclable package material have been implemented step by step. In Singapore, foods, paper, plastics, construction and demolition wastes, timber and gardening wastes, metals, residues, glass, textiles, and tires are recycled for domestic use or export. All unrecoverable combustible wastes will be collected

**Table 2.13** FSC certified area of forest

| Country | Quantity of certificates | Area of certified forest/$hm^2$ |
|---|---|---|
| Indonesia | 9 | 1,090,062 |
| Laos | 1 | 12,452 |
| Malaysia | 5 | 203,842 |
| Thailand | 4 | 7643 |
| Vietnam | 1 | 9782 |
| Total | 20 | 1,323,781 |

*Source* Fourth ASEAN State of the Environment Report 2009

and delivered to be burned in wastes-based power generation plants, while all unrecoverable and incombustible wastes will be delivered to Pulau Semakau Offshore Refuse Landfill.

Usually, 3R materials include waste paper, plastic wastes, scrap tires, glass, timber, nonferrous and ferrous metals, and building rubbish. Some ASEAN Member States have always separated and recovered material sources as a matter of practice, exercised by small enterprises and individual garbage pickers including women and children. Recycling in ASEAN is quite selective, driven by the market. The disposal of the un-selected recoverable wastes remains a big problem for many ASEAN Member States.

Informal separation or pickup of wastes takes place:

- At the source of wastes: separated or picked before the arrival of the collection vehicle;
- During collection: separated by collection staff during the loading; or
- At a disposal site: recoverable materials are collected by garbage pickers at a landfill or an open dump.

Garbage pickup can hardly provide a stable income source, although it is a daily activity of some communities in the ASEAN region. Therefore, some ASEAN Member States provide garbage pickers with financial and technical assistance with the aim to help improve the methods for collecting recoverable wastes and thereby increase their income. Meanwhile, they have provided measures to temper threats to the health and safety of garbage pickers.

Waste recycling will bring forth more job opportunities while easing the government's financial burden for solid wastes management.

#### 2.2.7.3 Trade of Environmental Goods and Services

Many international forums such as *Millennium Declaration, Monterrey Consensus, Doha Ministerial Declaration* as well as *Plan of Implementation of the World Summit on Sustainable Development* have presented discussions on the trade of environmental goods and services (EGS). Their commitments are helpful in supporting EGS liberalization and market expansion, which can serve as an important strategy favoring sustainable development. It would be helpful for ASEAN to tackle challenges and opportunities with regard to the EGS trade in the process of building itself into an efficient global economic zone.

A separated section is established for EGS in the negotiation tasks approved at the 4th WTO Ministerial Conference held in November 2001. The improvement in EGS availability and use will undoubtedly bring about numerous benefits, including reducing air and water pollution, raising the efficiency of energy and resources, and promoting the treatment of solid wastes.

It is predicted that the gradual trade liberalization and prudently controlled market of these industries will create job opportunities vigorously and promote the transfer of precious skills and techniques and thereby promote the economic growth of ASEAN indirectly. In short, the properly managed liberalization of EGS trade will help to promote the realization of sustainable development in the ASEAN region.

## 2.3 Comparative Results on China-ASEAN Green Development

The previous facts show that both China and ASEAN are making constant efforts to boost green development. However, the level of green development differs between China and ASEAN, especially among ASEAN Member States.

(1) China and ASEAN are both honoring their green development commitment with various efforts; however, due to the differences between economic development and natural resource endowment, their green development process and priority is also different.

- China's green development process started quite early. Situations are different among ASEAN countries. Several decades ago, attention to water and air pollution and solid waste treatment was very limited due to need for poverty reduction. As developing countries, the governments of many countries in Southeast Asia were inclined to boost GDP increase at the expense of environment and natural resources. Short-term trade-offs do exist between environment protection and business costs; however, lack of proper management in this regard will lead to the loss of competitiveness for economies in the region in the long run.
- Priorities of different ASEAN countries for green development are different. For example, Brunei is a high-income country, and its attention for environmental issue is largely going to global environmental issues and emerging issues. Due to shortage of water resources, Brunei is very active in sustainable management of water resources and achieved good results. Cambodia is one of the least developed countries but with very rich natural resources. Agriculture is its pillar industry, and its industrial base is very weak. Its environment suffered great damage during years of civil war, and the establishment of large number of new timber factories and coastal marine shrimp farming also led to the overdevelopment of forests.

(2) It is an arduous and multi-aspect challenge for both China and ASEAN to manage the natural resource endowment through maintaining a balance between protection and exploitation.

- Compared with developed countries, the resource utilization efficiency is still quite low in China and ASEAN Member States. Relative to population and economy, China's natural resources are not sufficient. Though ASEAN economies enjoy rich natural resources (forest, freshwater, marine resources, and rich biodiversity) on the whole, appropriate resource management has turned out to be top priority.

- In ASEAN Member States, their forest, river, and marine resources are encountering pressure from unsustainable practice. Their freshwater system is degrading. Without timely and bold measures, many areas in this region will suffer clean water shortage by 2030. Their renewable energy sector is faced with technological and political challenges. In the future, the demand for water resource in China will continue to rise. The spatial-temporal difference will be even more obvious, and the supply–demand conflict will be serious.

(3) China and ASEAN Member States all made certain progress in economic growth and pollution control, but due to increasing population, urbanization and the influence from such factors as industrial production activities, they are also are generally faced with severe air and water pollution.
ASEAN is a region with the most dynamic economies. However, fast economic development and inadequate attention to environmental protection led to serious ecological damages (see Figs. 2.37 and 2.38). Fast expanding urban centers posed challenges for the government to develop, implement, and maintain "green" strategies and policies. Urbanization usually goes hand in hand with intense industrial production in limited areas and dense population, all of which are the culprits of air and water pollution and people's health issue. Pollution will also increase the cost of production.

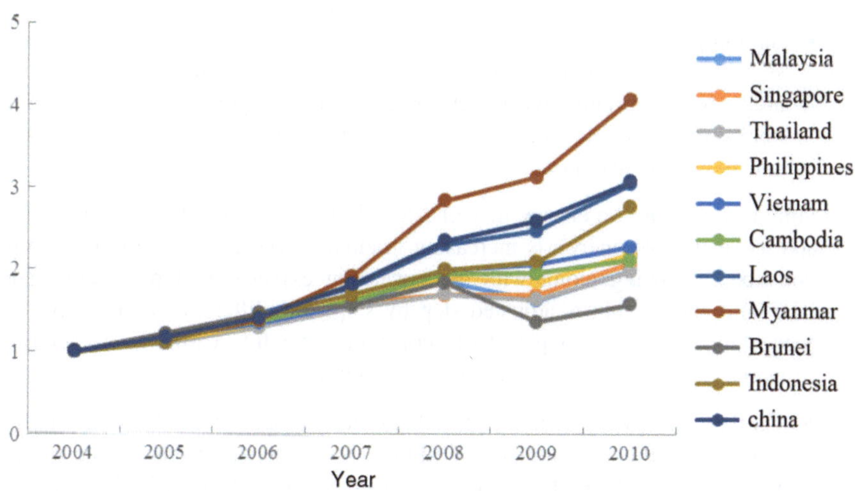


**Fig. 2.37** Comparison between China and ASEAN Member States by GDP growth. *Note* Growth of GDP for the period of observation is indicated by the value over the years divided by 2004 value

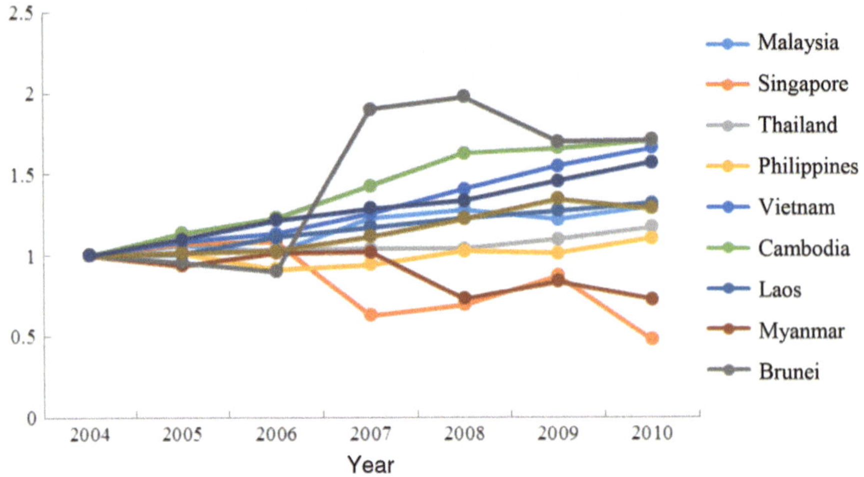


**Fig. 2.38** Comparison between China and ASEAN Member States by $SO_2$ emission growth. *Note* Growth of $SO_2$ for the period of observation is indicated by the value over the years divided by 2004 value

- The megacities of ASEAN are dramatically different in air quality. Many ASEAN cities are shrouded in smog—most is health-threatening particulate matter. In certain situations, smog has turned out a regional issue for Indonesia, Malaysia, and Singapore. Large-scale forest or agriculture land burning in order to open up new plantation area caused cross-border haze pollution, in particular in Sumatra Island and Borneo Island. Though China and ASEAN Member States reached consensus on controlling air quality, different countries have different policy measure priorities.
- Inappropriate solid waste treatment exacerbated the issue of clean water supply and air quality and triggered large-scale drainage issue in urban areas. Inappropriate environment practice further aggravated the burden of poor population, most of whom are exposed to unsafe drinking water, flood, and means of livelihood.

(4) Similar to China, the energy mix of ASEAN is also focused on fossil energy, and its renewable energy is increasing gradually. The gross energy consumption keeps growing along with the unceasing expansion of population, and GDP and energy mix is adjusted step by step. The difference is that ASEAN energy mix is based on petroleum consumption while China's coal-focused structure can be hardly changed in a short term.

- China's total energy consumption kept increasing with outstanding supply–demand conflicts. Though coal still takes the dominating share in the energy mix, the percentage of fossil fuel gets further lowered. In ASEAN countries, due to more coal and natural gas use, the percentage of fossil fuel in the

energy mix increased from 55.7% in 1990 to 72.4% in 2007 and the share of petroleum keeps the same at between 36 and 38%.
- ASEAN Member States pay special attention to the issue of climate change vulnerability and its impact as countries in Southeast Asia are especially prone to the impact of climate change. In the past years, ASEAN has taken various moves to address climate change, and climate change adaptation and mitigation have become important means to fight climate change.

(5) ASEAN is among the regions with the densest and most extensive forest resources, with a regional average coverage rate of 47%, much higher than that of China, 21.63%. However, the forest area is decreasing in several ASEAN Member States, along with the increasing population, the growth of agricultural production, logging and mining.

- The rich forest resource of ASEAN not only shaped its rich timber reserve, regional biodiversity with rich carbon resources, but also played important role in improving the ecological environment and climate in the region. Malaysia is one of the largest timber exporters in the world.
- For individual ASEAN Member States, Malaysia has long been short of the method for sustainable development, and its pollution management started quite late which led to the conflict of environment sustainability and economic development. Indonesia's social transition caused a serious of issues including environment deterioration, serious air, freshwater and marine pollution, coral reef degradation, and deforestation of primitive forest. In Thailand, urbanization incurred greater environmental pressure. Wild life and their living environment are threatened. In the Philippines, population pressure and damages of natural disasters lead to many environmental issues, e.g. shrinking forest area, land erosion, water and air pollution, large-scale destruction of coastal mangrove and wetland, and the damage of coral reef. The shrinking forest area in ASEAN region is mainly a result of increasing population, more agricultural production activities, lumbering and mining as many ASEAN countries are still relying on timber and agriculture export for foreign exchange earnings.

(6) For China and ASEAN, the greatest strains for sustainable consumption are attributable to the increasing emissions of wastes.

- Management of wastes has been a major challenge for China and ASEAN. Urban areas witness a big rate of wastes generation, which is aggravated further by other environmental problems arising from there.
- The most common waste management approach in China is incineration and landfill, and China is practicing strict and scientific water resource management system and encourages circular economy. Most of ASEAN Member States treat urban wastes with sanitary landfill or air storage and compost and incineration are also being adopted gradually.
- Reduction, Recycle, and Reuse (3R) of the wastes is getting popular in ASEAN. Most of ASEAN Member States have started different plants to

encourage 3R, including awareness raising and encouraging local communities to participate in waste management.

- ASEAN Member States adopted the concept of sustainable consumption and production which will promote a green economic transition without sabotaging their competitiveness. In contrast, China's overall consumption level is quite low, and people's awareness of environmental protection and green consumption has not been in shape yet.

## References


1. National Bureau of Statistics: China Statistical Yearbook 2002–2015. China Statistics Press, Beijing
2. Ministry of Environmental Protection: Annual Statistic Report on Environment in China 2001–2014. China Statistics Press; China Environment Press, Beijing
3. Olivier, J.G.J., Janssens-Maenhout, G., Muntean, M., Peters, J.A.H.W.: Trends in Global $CO_2$ Emissions: 2015 Report (2015)
4. Zhang, S.: Chinese scientists did first China-ASEAN regional ecological benchmark survey. http://www.chinageoss.org/gee/2014/D/D1/D1_1/. 2015-06-05/2015-11-12
5. National Remote Sensing Center of China: 2014 Annual Report on Remote Sensing Monitoring of Global Ecosystem and Environment (Eco-environmental Status in China-ASEAN Region). (2015)
6. China-ASEAN Environmental Cooperation Center: China-ASEAN Environmental Cooperation—Biodiversity and Regional Green Development. China Environment Press, Beijing (2013)
7. China-ASEAN Environmental Cooperation Center, Environmental Protection Bureau of Guangxi Zhuang Autonomous Region: ASEAN-China Environmental Cooperation: Innovation for Green Development. China Environmental Science Press, Beijing (2012)
8. China-ASEAN Environmental Cooperation Center: China-ASEAN Environmental Cooperation—Building up Partnership for Regional Green Transformation. China Environment Press, Beijing (2014)
9. Asian Development Bank Institute: ASEAN 2030. (2014)
10. Peng, B., Liu, X., Yang, Z.: Status of Resources and Environment and Potential of Cooperation in ASEAN. Social Sciences Academic Press, Beijing (2013)
11. ASEAN Secretariat: Fourth ASEAN State of the Environment Report 2009. (2014)
12. ASEAN Secretariat: ASEAN Documents Series 2011. Jakarta (2012)
13. "Green" energy development in ASEAN faced "Brown" barriers. Guangxi Energy Conserv. (3), 37–38 (2015)
14. ASEAN Centre for Biodiversity: ASEAN Biodiversity Outlook. Dolmar Press Inc. (2010)
15. China-ASEAN Environmental Cooperation Center: Progress of Implementing National Biodiversity Strategy and Actions Plans in China and ASEAN. China Environment Press, Beijing (2016)
16. APFP-SEApeat: Enhancing Sustainability of Forestry Practices on Peatlands. Global Environment Centre, Kuala Lumpur (2014)

# Chapter 3
# Drivers, Features, and Opportunities for Regional Green Development

**Abstract** Urbanization and industrialization have become the major drivers behind economic growth in China and the ASEAN region. The traditional extensive development pattern in the past can no longer sustain. Resource and environment constraints require China and ASEAN to leverage on the reform at supply and demand sides to promote green development. With very good ecological foundation and increasingly higher attention to environment protection, China and ASEAN have shaped fairly full-fledged ecological and environmental governance system. This chapter analyzes the environment-development relationship in China, Singapore, Vietnam, and Myanmar with the consideration of their economic development level and representativeness. It also analyzes the two major contradictions in green development: that between fast increasing economic aggregate and total emission as well as relationship of economic development and urban environmental capacity and quality during fast urbanization. The two sets of relations can affect green development. In addition, industrial transfer in China and the ASEAN region as well as such policy initiatives as Belt and Road Initiative also provided cooperation and development opportunities for regional green development.

## 3.1 Drivers for Regional Green Development

Economic growth in the era after the industrial revolution has been highly dependent on the use of fossil energy and resources. Unsustainable as it is, this mode of development has triggered a series of global and regional ecological and environmental problems that seriously threaten the health of residents and quality of life. Resources and the environment have become unable to support economic and social development, with ecological red lines repeatedly crossed. Against this backdrop, green development is not only an inevitable choice in the face of resource and environmental constraints, but also a promising direction of future economic and social development. Regardless of developed countries or developing countries, a transition to green economy is inevitable and imperative for building a more sustainable future. In the context of multiple global and regional environmental

 
China-ASEAN Environmental Cooperation Center, *China-ASEAN Environment Outlook 1 (CAEO-1)*, https://doi.org/10.1007/978-981-10-6211-7_3

challenges, green development and green transition has become the prevailing trend. China and the ASEAN Member States have developed green economy strategies, policies, and actions to accelerate the transition of economic development toward a greener future.

Similar to the global trend, green development in the ASEAN region is driven by forces from supply side and demand side. The former includes labor, land, capital, and innovation, while those of the latter include investment, export, and consumption.

Reform on the supply side is the proactive choice to adapt to the complicated international situation after international financial crisis. It is important for economic sustainability and is the must choice for the economy in a world with low growth, low demand, low inflation as well as high unemployment, high debt, and high bubble. The driving force on supply side is mainly to adjust land, capital, labor, and innovation with new technology, new products, new type of operation, and new business model. At current stage, traditional land use model is restricted by resources. The traditional financing model is no longer able to deal with the current market-based pattern. Global population aging has huge impact on economic growth and will also place great pressure on the world, in particular ASEAN region. Therefore, innovation has become the important driving force for future economic development and productivity. Science and technology are the primary productive force. Its innovation is a key driver for China and ASEAN to realize economic sustainability. As China's manufacturing industry has always been positioned in the low end of the international labor division, more efforts are needed to move it up in the industrial chain to adapt to the "new normal" state and foster new growth engines. Such measures will help release China-ASEAN regional growth potential and cross over the middle-income trap.

At the demand side, traditional drivers of investment, export, and consumption are facing multiple constraints including environment and resources. Mismatch between supply and demand exists, as a result of many ineffective and distorted demands. Therefore, guidance on green consumption should be provided to correct such mismatch. It means that people's higher income as a result of the economic development has changed the quality and form of consumption. Green, environmentally friendly, and healthy consumption model is regulating and shaping people's lifestyle which may affect supply side and change the consumption concept of the whole society. China-ASEAN Free Trade Zone has smoothened cross-border trade and investment. In terms of the investment home and abroad, the social requirement for green development and environmental protection is conducive to the fostering of new economic development models, e.g., green development, circular economy, and low carbon energy. In terms of export, as the economic development level and green development stage differ among countries in the region, domestic industrial restoration and transfer can narrow the wealth gap between these countries, while regional industrial production model shift and product innovation can adapt to the requirements of green economy and sustainable development. Both will further promote the industrial structuring and upgrading of the region, and facilitate energy conservation and emission reduction and optimized

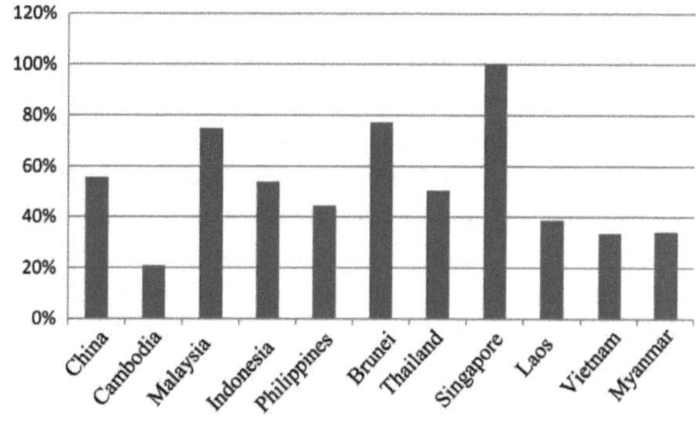


**Fig. 3.1** Urbanization rate of China and ASEAN in 2015 (World Bank database)

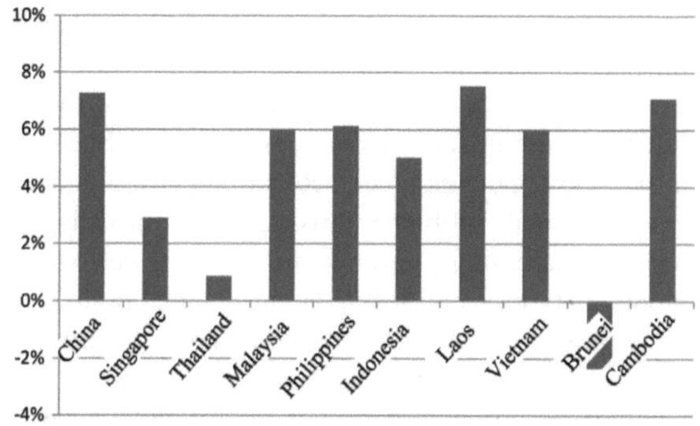


**Fig. 3.2** Economic growth rate of China and ASEAN (World Bank database)

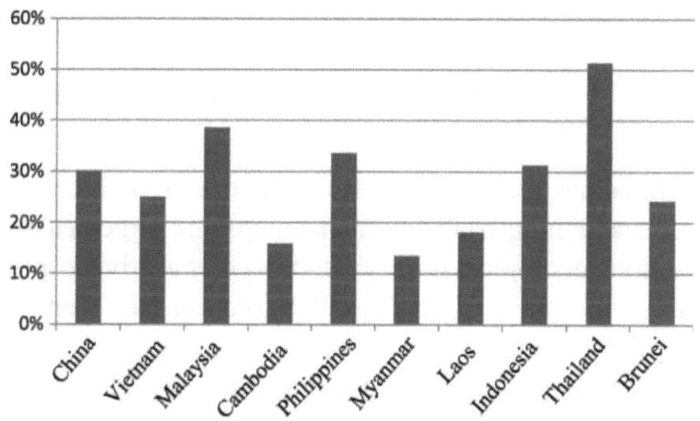


**Fig. 3.3** Higher education enrollment rate of China and ASEAN in 2013 (World Bank database) (2011 and 2012 data are used with 2013 data missing for Cambodia and Myanmar)

resource allocation, to further expand domestic and international markets for greater demands.

In this process, urbanization is the important driving force (see Fig. 3.1). The urbanization rate is rising in ASEAN Member States which greatly spurred economic growth (see Fig. 3.2). Other factors including labor quality have also promoted economic development and social transition (see Fig. 3.3).

The relations between environment and development in China and ASEAN Member States are diversified by economic and social development stages and resource endowments. Hence, the corresponding social, economic, and environmental conflicts, and the priority order are different, whether they are purely environmental or involve the relationship between environment and development. The traditional mode of growth entails high carbon emissions, high resource consumption, and serious environmental damage and causes imbalances in economic, social, and regional development. A common consequence for countries with different national conditions is environmental pollution and ecological damage. In view of the low added value, waste of resources, and environmental destruction, it is necessary to shake off the reliance on high consumption of fossil fuels and resources and the consequent high pollution, in order to achieve resource conservation and environmental improvement while pursuing economic growth with low emissions. Typically, China's manufacturing sector, situated in the front end and low end of the international industrial chain, lacks core values and independent innovation capability and brings great pressure on the environment and ecology. Transformation is urgently needed to remove the major obstacle to the country's future economic and social development. Such problems should be addressed through green development. Regarding ASEAN Member States, the ecological and environmental problems arising from urbanization and industrialization also become increasingly prominent, including serious damage to grassland and forest, biodiversity loss, and severe environmental pollution. It calls for green development to reduce resource consumption, minimize environmental damage, or decouple economic growth from resource consumption and environmental pollution. In addition, green development is a very effective way to truly protect biodiversity[1] and will give rise to new economic growth points.

## 3.2 Features of Regional Environment

China and the ASEAN region have a favorable ecological baseline, but urbanization and industrialization that involves large-scale use of resources have caused serious ecological damage and even exacerbated poverty. However, on the whole, the importance of ecological protection has been generally recognized within the region.

It is clearly understood that green development is a most effective approach toward sustainability.

[1]Biodiversity, the material basis for the survival of mankind, has direct value and indirect value. The direct value includes food, medicines, and industrial raw materials that human obtains from wild and domesticated components of biodiversity, as well as recreation and tourism value. The indirect value is mainly associated with the functions of ecosystems, covering hydrological process regulation, soil erosion prevention, climate regulation, pollutant absorption and decomposition, nutrient storage and recycling. In the long run, the maximum value of biodiversity is to provide human with the living conditions for adaptation to local and global changes. Green development emphasizes sustainability of social and economic operation and fundamentally protects biodiversity and ensures normal ecosystem services by minimizing damage to the ecological environment in development process.

### 3.2.1 Intra-regional Overview

The environment and development relationship is closely linked to the stage of development. The industrialization process is an important measure of the level of economic and social development of a country or region. In China and the ASEAN region, the environmental problems have become increasingly apparent over the years, such as ecosystem and biodiversity degradation (deforestation, forest fires, and arable land reduction), shortage of resources (water shortage and food security), environmental pollution (air pollution, water pollution, and waste pollution), and frequent natural disasters caused by global climate change (typhoons and storms). In recent years, however, with the decline in the growth of industrialization, the abundance of natural resources gradually improves amid a slowdown in the economy. While the level of environmental governance is subject to economic and technological constraints, the degree of environmental pollution and damage, affected by awareness of environmental protection, varies sharply among ASEAN Member States.

From the perspective of economic development, as measured by gross domestic product (GDP) per capita, Singapore and Brunei rank front, followed by Malaysia, Thailand, the Philippines, Indonesia, and then are Vietnam, Cambodia, Myanmar, and Laos. In 1990, the average GDP per capita of the six ASEAN Member States was 11 times that of these last four countries, and in 2012, the difference was narrowed to 3.5 times [1]. According to the International Monetary Fund (IMF), in 2013, the GDP per capita reached 52,179 US dollars in Singapore, making the country one of the world's richest, and the number was several times smaller and only read 884 US dollars in Myanmar. In 2014, the per capita GDP in China, Brunei, Cambodia, Indonesia, Laos, Malaysia, Myanmar, the Philippines, Singapore, Thailand, and Vietnam were 7593.9, 40,776.3, 1084.4, 3514.6, 1707.5, 10,829.9, 1197.5, 2843.1, 56,286.8, 5560.7, and 2052.3 US dollars, respectively (current US dollars, the World Bank). Many areas in Cambodia, Laos, Myanmar, and Vietnam remained stuck in poverty even though the ASEAN has gained overall fast economic development. In 2014, more than 30% of the Laos population, in 2011, 16% of the Indonesian population, and in 2010, 10% of the Cambodian population lived below the poverty line (1.25 US dollars per day, the World Bank). By attracting international investment and developing import substitution industries, Indonesia, Thailand, Singapore, the Philippines, and Malaysia have grown fast into moderate economies since their reform and opening up in the 1970s, creating the "Southeast Asian Miracle". In the face of global economic integration, the ASEAN FTA advances at a quicker pace with the inclusion of Myanmar. Unable to free from the negative impact of the slowdown in the EU, US and Japanese economies, the ASEAN Member States gradually diversify exports and increase interconnection with such emerging economies as China and South Korea. At a different stage of development from ASEAN Member States, China has a low level of per capita GDP and a dual economic structure with strong manufacturing industry and developed tertiary industry. China has both cooperation and competition with the ASEAN Member States.

From the perspective of environmental resources, there are quite a few differences in resource consumption and pollutant emissions between China and the ASEAN Member States. According to the integrated resource and environmental performance assessment carried out by the Chinese Academy of Sciences (CAS), China's Environmental Performance Index was 4.98 in 2012, while the index read 1.75 for Indonesia, 1.69 for Malaysia, 0.57 for Singapore, and 2.30 for Thailand (the assessment set the benchmark of 1 for year 1990 and considered seven categories of indicators on resources and pollutants). Natural resource endowments are great in China and the ASEAN region, but significant differences are still observed in geographic conditions, climatic conditions, mineral resources, and energy stocks.

ASEAN is the world's main producer of rice and a world-famous producer of natural rubber, coconut, sugarcane, and its products also include copper, tungsten, manganese, and precious stones. All ASEAN Member States, except Laos, have coasts and abundant marine resources, and Malaysia, Indonesia, and Brunei are rich in natural gas. Brunei and Malaysia rank front in both ecological resources per capita and economic development, while ecological resources per capita are limited in other countries.

The level of environmental governance also varies in China and the ASEAN Member States. It is directly related to the level of economic development and the degree of attention to the environment. On the whole, a relatively complete and increasingly sound environmental governance system has been put in place in these countries.

Nevertheless, forests and rivers suffered destruction under the previous extensive organizational structure. In 2008, for example, 54% of 33 monitored rivers in Indonesia were seriously polluted, and 14–28% of the rivers in the Philippines exceeded the limit of biological oxygen demand (BOD). Singapore sets a global model for greening the country, and the forest coverage reaches 89% in Laos, but only 37% in Thailand [1].

### 3.2.2 Green Development

The concept of a global green economy was introduced by the United Nations Environment Programme (UNEP) in 2008 to cope with the global economic recession. Ever since, green development has been gradually accepted and widely acknowledged worldwide.

The Chinese Government attaches great importance to environmental protection and identifies environmental protection as a basic national policy to promote sustainable development. With increased input over the years, China has made very outstanding achievements in global and regional environmental governance, noticeably climate change mitigation, atmospheric haze pollution control, and transnational river protection. Nevertheless, the contradictions remain very sharp between economic development and the environment in the rapid process of industrialization and urbanization, giving rise to prominent pollution problems associated with the interests of residents. To this end, stressing the need to respect nature, the Chinese Government has incorporated ecological progress into the whole process of advancing economic, political, cultural, and social progress in the socialist cause with Chinese characteristics.

"Greening" was first mentioned and put in same place as new-type industrialization, urbanization, information technology, and agricultural modernization in the opinions on promoting the Ecological Progress. Green ideas and values are incorporated into national development to green the ways of production, life, and consumption. These encompass, on the enterprise level, green development modes, green industries, and green products, and on the government level, green management, green governance, and green education. Green development identifies environmental resources as an integral factor in the socioeconomic development and an indicator of economic, social, and environmental sustainability. Greening provides an opportunity for development under the new normal and outlines the direction, objective, and criteria for future economic development.

China's 11th Five-Year Plan (2006–2010) put constraints on energy consumption per unit of GDP and total pollutant emissions for economic development and set the targets of 20 and 10% reductions respectively in the accountability system. In these five years, chemical oxygen demand (COD) and sulfur dioxide ($SO_2$) emissions fell by 12.45 and 14.29%, respectively; carbon dioxide ($CO_2$) emissions decreased by 1.46 billion tons, reversing the increasing trend of total emissions in the previous years (2000–2005). The 12th Five-Year Plan (2016–2020) set down seven binding targets on energy and environment: a 17% reduction in carbon emissions per unit of GDP, a 16% reduction in energy consumption per unit of GDP, raised share of non-fossil energy to 11.4%, a 8% reduction in $SO_2$ missions, a 8% reduction in COD emissions, and a 10% reduction in ammonia nitrogen and nitrogen oxide ($NO_x$) emissions. The rest targets included cumulative energy savings of 670 million tons of standard coal, a 10% decline in each of the COD and $NO_x$ emissions, and total energy consumption within 4 billion tons of standard coal over the five years of 12th FYP. In comparison, these new indicators are more stringent and decomposed and assigned from higher to lower levels.

At the 2009 Copenhagen Summit, the Chinese Government committed to, by 2020, cutting the $CO_2$ emissions per unit of GDP by 40–45% from 2005 levels and increasing the share of non-fossil fuels in primary energy consumption to around 15%. In the China-US Joint Announcement on Climate Change signed in November 2014, China planned to peak the $CO_2$ emissions around 2030 or early and further raise the proportion of non-fossil fuels in primary energy consumption to around 20%. In June 2015, Chinese Premier Li Keqiang committed to (a) peaking the $CO_2$ emissions around 2030, with intention of trying to peak early; (b) cutting carbon intensity by 60–65% of 2005 levels; (c) raising the share of non-fossil fuels in primary energy consumption to about 20%; and (d) increasing the forest stock by about 4.5 billion $m^3$ from 2005 levels.

In the context of nationwide fog and haze pollution, a number of provinces, municipalities, and autonomous regions have included respirable particulate matter ($PM_{2.5}$) as an important control indicator. Clearly recognizing the importance of sustainable development, the Chinese Government once again stressed "intended nationally determined contribution" at the Paris Climate Change Conference in 2015. It has allocated 20 billion Chinese yuan to the South–South Climate Cooperation Fund to support international cooperation on clean energy, disaster prevention and mitigation, ecological protection, climate-adapted agriculture, and intelligent low-carbon cities. The 13th FYP period (2016–2020) will be important for China to achieve a transition to green economy and comprehensively improve environmental quality.

The ASEAN Member States have also recognized the importance of green transition, which is reflected in the ASEAN Socio-Cultural Community Blueprint (2009–2015) and the Manila Declaration on the ASEAN Environment (1981). They are also committed to green development, with an aim to build a resource-saving and environment-friendly society. As far as new energy is considered, the Philippines plans to invest 9 billion–10 billion US dollars in new energy projects during 2009–2019 and aims to double power generation from renewable energy sources. Indonesia focuses the efforts on alternatives to fossil fuels, including biofuels, and has tried to cut the use of fossil fuels by 20% every year from 2005. Malaysia has developed the National Energy Policy (1979), National Depletion Policy (1980) and Fuel Diversification Policies (1981, 1999). Thailand introduces tax breaks to stimulate the development of bio-diesel, bio-ethanol, and solar energy. Singapore plans to increase energy efficiency by 35% and recovery rate to 70% by 2030. Malaysia and Vietnam have already worked out the roadmaps for environmental governance and cleaner production. In the post-Rio+20 era, the ASEAN Member States have the advantage of late development and can carry out exchanges on environmental protection technologies and experience, through cooperation between the government and the private sector.

Green transition makes China and ASEAN more closely connected. It favors greenhouse gas (GHG) emissions reduction, pollution control, and efficient use of natural resources, and promotes the development of green industries. In the transition to a green economy, China and ASEAN have a lot of room for cooperation and improvement in sewage treatment, air pollution control, cleaner production technologies, and new energy development. Yet, the focus of environmental protection is not identical among countries, considering the national characteristics. China's environmental pollution problems are complicated by the vast land area and sharp geographical differences, and can be deemed as a comprehensive combination of pollution problems of the ASEAN Member States. China's environmental policy puts the focus on pollution prevention and control and environmental quality improvement. In ASEAN, environmental governance performs as a result of increasing environmental awareness and deteriorating environment. In addition, the rise of environmental non-governmental organizations (NGOs) challenges the traditional model of country-driven environmental governance and excites the attention of developing countries to green development. With the common goal of green development, China and ASEAN can make use of their complementary advantages in resources.

### *3.2.3 Environment and Development Relations*

There is a unity of opposites between sustainable economic development and environmental protection. It is not appropriate to ignore the environment when pursuing the economic development or ignore the economy while protecting the environment. To date, the resources required for industrialization have been far beyond what can be provided by the nature, while the restoration rate of the natural environment falls short of the pace of economic development. Irrational economic structure and economic order are bound to lead to ecosystem imbalances and natural resource depletion as a result of soil erosion, overgrazing, and deforestation. Despite the economic accumulation, the restoration and improvement of the natural environment will take time. With given productivity, pollution control and environmental protection require technical and financial support, which, to some extent, hinders economic progress. Such trade-offs need to be considered in dealing with the relationship between economic development and environmental protection.

China and ASEAN have accomplished remarkable economic achievements. Noticeably, since the reform and opening in 1978, China has maintained an average annual GDP growth of nearly 10%, far higher than the world average during the same period. However, the economic progress entails high environmental costs, manifested in emissions of major pollutants beyond the environmental carrying capacity, common water, air, and soil pollution, and prominent solid waste, car exhaust, and persistent organic pollutants (POPs). ASEAN encounters a concentrated breakout of environmental problems that usually arise from centuries of industrialization in developed countries. In China, the country-wide fog and haze pollution in recent years makes into the social and environmental focus. Environment has become a major livelihood issue, and the pressure mounts with economic growth. The near future will see not only important opportunities for economic and social development, but also more apparent resource-environment contradictions. It is imperative to put environmental protection in the same position as economic development for coordinated progress and to protect the environment in an all-round way which is not limited to administrative measures. It should be recognized that environmental protection should be incorporated into economic development. The ASEAN countries (excluding Singapore) fall into the dilemma between economic priority and environmental protection, and the dysfunctional traditional model of country-driven environmental governance even adds difficulty in dealing with the relationship.

Poverty is another issue that cannot be ignored. The poverty-stricken areas generally suffer noticeable ecological fragility and environmental pollution. The majority of the poor also bear the risk brought by ecological damage. The environment issue, viewed as a development issue, can be eased through sustainable development that pursues harmony between man and nature. Given the fact of poverty in the region, proper solutions to poverty alleviation and environmental protection will also be an important part of green development.

In this context, China and ASEAN Member States gradually deepen their understanding of green economy and realize that protecting the ecological environment is protecting productive forces, and improving the ecological environment is developing the productive forces. The new climate economics advocate the development of a low-carbon economy to promote renewable energy and industrial waste disposal. It is a move to deal with climate change and to stimulate the economy with new growth poles by adjusting industrial structure and developing emerging industries. Fast-growing cities and regions also pursue low-carbon green growth and encourage environmental protection industries. In light of this, measures reflecting national conditions are taken in countries.

### 3.2.4 Institutional Reform

Institutional reform provides an important guarantee for economic development and favorable environment. A sound resource price formation mechanism that reflects such external costs as resource scarcity, market supply and demand, and environmental damage can drive the transformation of economic development and promote a favorable environment. It is also possible to solve deep-seated contradictions and problems of economic development. This requires transformation of government functions and institutional reform to make up for the absence, translocation, and dislocation of government's role in economic development. The costs of land, mineral, and water resources should be reflected in the prices of production factors adjusted through the market mechanism. Governments need to strengthen social regulation by improving laws and regulations on resource conservation, environmental protection, and ecological security, in order to change problems arising from the blind pursuit of economic development. In China, for example, under the current GDP-based performance evaluation system, local governments pursue excessive investment to drive economic development. Production factor prices form with little involvement of the market, and resources allocation is quite subject to administrative management.

Institutional reform is the most fundamental and far-reaching way of promoting economic development and favorable environment. In order to effectively protect the ecological environment, China and ASEAN Member States have launched adjustments to institutional frameworks while pursuing economic growth, in line with their stage of economic development and the political system. For example, China identifies knowledge and technology intensive industries as the core of development in the middle and late stage of industrialization. It has entered the era of integrated reform for promoting ecological progress. Environment related institutional frameworks cover environmental protection, energy conservation, and emissions reduction, climate change, circular economy, resource conservation, and ecological protection.

The current environmental protection situation is grim. New problems continue to emerge, while the old ones remain unresolved. The quality of the environment is far below expected by residents. To this end, the reform for environmental protection has become a necessity, which encompasses scientific and rational top-level design, overall deployment of organizations, and management reform. A good ecological environment itself is productivity and core competency. Through management reforms, the modes of economic development are transformed and upgraded, and production and consumption patterns changed, to drive economic restructuring and upgrading and optimize the economic development. Meanwhile, energy and environmental criteria and environmental access thresholds are also raised, leading the development of new industries, to produce environmental, economic, and social benefits.

The major global technological revolutions, such as new energy revolution and Internet Plus, will bring more opportunities for industry development and use the hand of the market to promote "green development". Economic, social, and environmental are the globally recognized three pillars of sustainable development. More ASEAN Member States will benefit from globalization if earlier and more deeply involved in. With low levels of economic development, the ASEAN Member States, except for Brunei and Singapore, can take the advantage of late development. They can summarize and draw on the experience of economically advanced countries and push ahead with environmental protection through technological and cultural cooperation in the field of environmental protection. It should be noted that the integration within ASEAN itself has sharpened the advantage of collaboration and build a firm foundation for South–South cooperation and North–

South cooperation. Meanwhile, the institutional reforms at different levels lay the foundation for sustainable development of ASEAN Member States.

## 3.3 Classification of Environment and Development Relations

China and most of the ASEAN Member States are mostly developing countries with economic disparity due to differences in historical, political, and natural conditions. To truly achieve green development, differentiated paths should be identified based on analysis of development status of different countries.

In view of economic imbalances, countries in the region are classified according to the relationship between environment and development. This requires an indicator system that fully considers the economic, social, and environmental dimensions. Based on evaluation using characteristic indicators, the priority order of environment and development problems is clarified specific to countries. Due to language difference and data availability, qualitative analysis is provided here.

### *3.3.1 Measurement Benchmark*

#### 3.3.1.1 Economic Development Measurements

The level of economic development refers to the size, speed, and standard that an economy has achieved. In the context of globalization, the service sector continues to be isolated from the manufacturing sector with the refined international division of labor, and the advantage of late development plays an increasing role. Industrialization has become an important indicator of economic development.

According to Petty-Clark's law and Chenery' and Kuznets' empirical analysis, the internationally recognized indicators to measure industrialization include per capita GDP, percentage of non-agricultural added value, percentage of non-agricultural employment, and urbanization rate (see Table 3.1). Based on this,

**Table 3.1** Criteria to measure the level of economic development

| Development stage | Per capita GDP/USD | Percentage of non-agricultural added value/% | Percentage of non-agricultural employment/% | Urbanization rate/% |
|---|---|---|---|---|
| Early-industrial stage | 1200 | 65 | 20 | 10 |
| Transitional stage | 2500 | 80 | 50 | 20 |
| Middle stage | 5000 | 90 | 70 | 60 |
| Post-industrial stage | 10,000 | 95 | 90 | 80 |

the process of industrialization is divided into early stage (infancy), transitional stage (from early stage to middle stage), middle stage (basic realization), and post-industrial stage (full realization). China and ASEAN Member States can be divided into four categories in accordance with the level of economic development, namely (a) post-industrial countries, such as Singapore; (b) middle-stage industrial countries, such as China; (c) transitional-stage industrial countries, such as Vietnam; and (d) early-industrial countries, such as Myanmar. Details are provided below.

#### 3.3.1.2 Social Development Measurements

Social development measurements are important criteria measuring the extent of social development and degree of harmony. Among them, the most important is the Human Development Index (HDI) developed by the United Nations Development Programme (UNDP) [2]. It considers income, education, life expectancy, and infant mortality and is calculated according to certain methods. In 2014, the world average of the index was 0.702, and HDI read 0.719 in China, 0.901 Singapore, 0.852 Brunei, 0.773 Malaysia, 0.722 Thailand, 0.64 Vietnam, 0.660 the Philippines, 0.586 Indonesia, 0.584 Cambodia, 0.569 Laos, and 0.524 Myanmar. Singapore ranked 1st among the ten countries and 9th in the world, while the Chinese mainland ranked 91st only. The rankings reflect national differences in relations between economic development and human development. For example, China was in the 134th position in the rankings in 1980, but HDI continued to rise slowly, up by 0.02–0.719 in 2014 from 0.699 in 2013. The improvement reflects the growth of per capita GDP and the great economic progress in China.

The growth rate of HDI is also an important indicator. China's HDI growth rate experienced a downward process, specifically slowdown following the sharp rise after changes on the basis of extreme poverty. HDI fully demonstrates that economic growth alone does not automatically and completely translates into human development which also involves education, nutrition, health, and the environment. As in China, HDI does not improve correspondently with GDP which grows at an annual rate of 10% over the 30 years since the reform and opening up. It is undermined by "negative products" of economic growth, such as environmental pollution and ecological damage and the threats to human health. Nevertheless, HDI rises in China and the ASEAN, at lightly different rates, which explains the national interaction and common development. In the global trend of economic integration, HDI and the growth rate will further change to varying degrees with increasingly close cooperation and trade in Southeast Asia, especially between China and ASEAN.

#### 3.3.1.3 Environmental Governance Measurements

ASEAN used to be recognized as a model of global environmental protection, but now, it suffers severe environmental degradation. It is prevalent in the region that environmental protection is outweighed by economic development. Moreover, the high demand of the international market for resource-intensive products intensifies the overall deterioration of the environment. Environmental governance is uneven across countries within the region, due to differences in resource endowments, socioeconomic development, political regimes, and cultural traditions. Increasingly recognizing the importance of environmental protection, these governments have taken measures in line with national conditions to address environmental problems, such as sharing environmental governance information, developing environmental evaluation criteria, implementing environmental performance audits, and evaluating government policy effects. With consideration to economic policies, efficiency and effectiveness, environmental governance measurements include effects of environmental policy and administrative controls, environmental benefits generated by environmental protection funds, efficiency of funds, and performance of environmental projects. Referring to and learning from the experience of Western developed countries, China and ASEAN gradually improve the government-led environmental governance system by encouraging the broad participation of businesses, general public, and NGOs.

Environmental governance in the region is subject to the specific conditions of countries. In addition to industrialization and urbanization, the application and management of clean energy and environmental technologies are important to pollution control. Economic development is also a major cause and influential factor. Economic development levels, if not decoupled from environmental pollution, are closely related to pollution severity. Regional environmental problems also affect environmental governance. For example, the special pollutants generated in Indonesia's forest fires affected not only Indonesia, but also Malaysia and Singapore. In some areas, the pollutants concentration far exceeded the acceptable level for humans, under the dual impact of global climate change and regional environmental pollution. River protection in ASEAN also requires participation of different countries.

Environmental governance in the region is also undermined by the lack of national and local policies and mechanisms. In this regard, a particularly prominent problem is cooperation on cross-border pollution control. Hence, the measurement of environmental governance considers results of cross-border environmental cooperation, in addition to policy improvement, development, and implementation in respective countries. Country classification is then developed from the above analysis.

### 3.3.2 Country Classification Results

China and ASEAN Member States are categorized according to the relationship between environment and development. Considering the level of economic development and representativeness of countries, only the major representative countries are selected and analyzed in this chapter, as shown in Table 3.2. The stages of development determine the measures and methods that should be taken in the future.

The basic situation of these four countries is specified below.

**Table 3.2** Classification of representative countries in China-ASEAN region

| Development stage | | Post-industrial stage | Middle stage | Transitional stage (early to middle stage) | Early-industrial stage |
|---|---|---|---|---|---|
| Representative nation | | Singapore | China | Vietnam | Myanmar |
| Characteristics of the environment-development relationship | Huge disparity within the region | Highly developed economy, good environment, scarce natural resources | Steady and rapid economic growth, continued eco-environmental deterioration, natural resource scarcity | Agriculture-based economy, serious environmental pollution, abundant natural resources | Agriculture-based underdeveloped economy, good environment, abundant natural resources |
| | Common green development goals | Singapore Green Plan, aimed at a healthy living environment and sustainable social development | Ecological Civilization Strategy, intended to reverse the trend of ecological deterioration from the source | Poor environmental work results, given imperfect environmental protection system and relaxed environmental quality standards | No clear action |
| | Environment-development relations to be measured | Little impact of economic growth on the environment in pursuing a circular economy and sustainable development | Acute environmental problems. Priority should be given to protecting the environment and build an environment-friendly society | Grave environmental problems, economic development given priority over environmental protection, economic growth at the cost of environmental damage | Priority given to economic development, no serious environmental problems |

(continued)

**Table 3.2** (continued)

| Development stage | | Post-industrial stage | Middle stage | Transitional stage (early to middle stage) | Early-industrial stage |
|---|---|---|---|---|---|
| Challenges facing green development | Demand for economic growth in a reality of environmental deterioration | Natural resources failing to meet the needs of economic development | Huge resources and environmental costs of economic growth, challenges in a transitional period | One-sided pursuit of economic development, ignorance of environmental protection | Focus placed on economic development, without giving consideration to resources and environmental protection |
| | Conflict between increased pollution load and limited environmental capacity | Good environment, but increasing pressure on environmental carrying capacity | Ecological damage aggravated by industrialization and urbanization | Lack of effective environmental management, increasing emissions of pollutants | Environmental pollution from agriculture serious emissions of small-and medium-sized enterprises (SEMs) |
| | Social development lagging behind environmental targets | Failure to reduce consumption of resources in the context of severe shortage of natural resources | Develop and undertake medium- and long-term environmental quality targets | Inappropriate methods and backward management, hindering environmental protection measures from producing the desired results | Environmental targets to be clarified |

#### 3.3.2.1 Huge Disparity Within the Region

Singapore is one of the Four Little Dragons of the Asian economy. It is highly dependent on the US, Japanese, and European economies and the neighboring markets. Adhering to a liberal economic policy, Singapore invests heavily in infrastructure and strives to attract foreign investment with the most favorable business environment. In recent years, the country actively promotes the "regional economic strategy" that accelerates overseas investment. Singapore embraces the world's third largest oil refining center, and sees fastest growth of the electronics sector among competition from other Asian countries. The economy, developing with sound momentum, mainly encompasses commercial, manufacturing, construction, financial, transportation, and communications sectors. Low-carbon industrial services rise as the leading economic growth point, while tourism contributes greatly to foreign exchange earnings.

China has maintained an average growth of 9.6% in the past three decades since economic reform and opening up. After addressing the subsistence problems, the country is moving forward toward a well-off society in an all-round way. Especially in the new century, under the premise of a substantial increase in the aggregate product, deflation is eliminated without obvious inflation. Economic growth becomes more coordinated and steady with improvement in quality and efficiency. By virtue of continued economic development, China's comprehensive national strength and people's living standards have markedly improved. Currently, the country ranks first in the world in terms of foreign exchange reserves. Progress is also seen in infrastructure construction, manufacturing capability and capacity, and independent innovation capability. Labor quality has improved significantly with the all-round development of science and technology education. As opening up expands, China becomes more closely connected with the world economy.

In Vietnam, agriculture dominates the economy, providing livelihood to 80% of the national population and contributing 40% of the GDP. Farmland and woodland cover 60% of the national total area. Rice remains the primary crop, while a variety of cash crops are grown, including fruit, coffee, rubber, cashew nuts, tea, peanuts, and silk. In recent years, Vietnam has risen to the world's second largest rice exporter and third largest coffee exporter. The industrial sector, contributing 24% of the GDP, mainly includes coal, power, metallurgy, and textiles. Highly valued by the Vietnamese Government, the oil industry is an important pillar of the national economy and a major source of foreign exchange earnings. More specifically, oil accounts for one-third of the national value of exports and one-quarter of foreign exchange earnings. In recent years, the processing industry has increased the production, but still develops slowly, due to low efficiency. Tourism has been greatly developed since inception in early 1990s owing to rich tourism resources.

Myanmar is listed in the world's least developed countries (LDCs) by the United Nations. Agriculture is fundamental to the national economy and contributes 60% of GDP. The main crops are rice, wheat, corn, peanut, sesame, cotton, beans, sugar cane, palm oil, tobacco, and jute. Animal husbandry and fisheries are mainly dominated by private business. Given extremely rich forests and fishery resources and a forest coverage rate of 57%, Myanmar is reputed for its forests. Among over 2300 tree species, teak accounts for 30% of the national forest area and 85% of the international market supply. Myanmar is also renowned for its jade and precious stones in the world. Industrial output accounts for about 10% of the GDP, and the dominant private sector contributes 75% of the GDP. Industries in the country mainly include small-scale machinery manufacturing and assembly, textile, printing and dyeing, rice milling, wood processing, oil and gas extraction, sugar, paper-making, fertilizer, and pharmaceuticals.

#### 3.3.2.2 Common Green Development Goals

Singapore as an independent state needs to overcome a variety of problems, such as high unemployment rate, housing shortage, inadequate infrastructure and communications equipment, and urban environmental degradation. Due to lack of natural resources, the country tries to maintain a steady growth by developing port facilities, providing affordable housing, promoting tourism, and accelerating industrialization for more job opportunities. However, this rapid development has exerted a negative impact on the environment. To this end, the government implements long-term planning for protection against environmental threats. To achieve comprehensive monitoring of environmental quality, through effective legislation, adequate facilities for environmental protection are put in place, combined with rational use of technical measures. Environmental control has been linked to territorial planning, in order to ensure development at appropriate sites and in harmony with the surrounding land use. Architectural planning and completed buildings must accept the examination of environmental protection departments. Based on online monitoring, the preventive control, and mandatory control of air and water pollution are strengthened. Hazardous chemicals and toxic substances are brought under strict control and proper management to ensure safety. Public sewage treatment systems have been established, covering residential, commercial, and industrial buildings. Engines and fuel quality are improved in an effort to reduce emissions, and vehicles and fuel consumption are controlled by way of traffic management. The country also implements daily centralized garbage collection and disposal and garbage minimization and recycling. Singapore Green Plan sets a new strategic direction toward sustainability of environment and development.

In China, grave environmental pollution problems accompany economic growth. As far as resources are concerned, the problems include rapid consumption, serious shortage, severe pollution, and low efficiency of utilization. Urban clusters are dense, and industrial development beyond environmental carrying capacity leads to acute air pollution and quick ecological degradation that threaten human health. In face of these problems, the Chinese Government focuses efforts on environmental protection, such as establishing sound environmental laws, regulations and standards, and taking measures for environmental protection. Economic instruments are placed in a prominent position, while legal instruments are deepened and improved. A series of micro-level mechanisms have been successively introduced, covering environmental labeling, pollution charge, water pollutant permit, emission trading, environmental impact assessment (EIA), shutdown of polluting enterprises, and corporate accountability for environmental targets.

At present, the Chinese Government is building an Ecological Civilization which fosters resource-saving and eco-friendly geographical pattern, industrial structure, and ways of production and life through green, circular, low-carbon development. It is a move to create a favorable production and living environment by reversing the trend of ecological environmental deterioration from the source and will make contribution to the global ecological security.

In Vietnam, economic development has stimulated the social demand for natural resources, resulting in exhaustion of natural resources, increased natural disasters, and soil erosion. In the long-term waste disposal process, soil, surface water, groundwater, and air have been polluted, undermining ecosystems and posing a direct threat to human health. Moreover, there is a trend of gradual aggravation of soil, water, air, and noise pollutions. Agricultural development accelerates the deterioration of environmental conditions by exerting negative impact on upland forest, coastal areas, and low-lying farmland. The Vietnamese Government has launched a series of initiatives to deal with environmental problems. However, the newly amended and enacted ambient air quality standards set less stringent requirements and relatively low thresholds. Regarding water pollution, the target is set to curbing the rising trend, in which production facilities that cause serious environmental pollution and wastes are restricted by pollution charges. The Vietnamese Government does environmental reporting on a regular basis instead of environmental work assessment. There is a general lack of data on the environment and other sectors. Currently, monitoring tasks, information collection, and database development do not meet regulatory requirements in many areas. Limitations are also seen in data exchange and information sharing, cross-sectoral cooperation, and environmental reporting.

Myanmar embraces an agreeable environment and abundant natural resources, particularly forests and fishery resources. However, serious deforestation has led to a 20% growth of the $CO_2$ emissions. Myanmar Agenda 21 fully embodies Myanmar's emphasis on sustainable development. In 2012, the Environmental Protection Law of Myanmar was promulgated, but still remains a general framework that does not provide implementation rules, EIA methods, and environmental standards. Infrastructure for environmental protection purpose needs construction and improvement. Under the cooperative framework of ASEAN, Myanmar continues to strengthen environmental awareness. Regional cooperation on environmental protection has been reflected at different levels of laws, such as China-ASEAN Strategy on Environmental Protection Cooperation.

#### 3.3.2.3 Environment-Development Relationship

Singapore adheres to circular and sustainable economic development, avoiding economic and industrial growth at the expense of the living environment. By way of appropriate planning and control and strict and scientific laws and regulations, the impact of urbanization, industrialization and economic behavior on the environment has been minimized.

China has paid a heavy environmental price for rapid economic growth. While air, water, and soil pollution remain quite serious, food safety emerges as a major environmental issue that threatens human health. China has entered a period of frequent environmental pollution accidents and has to deal with tremendous pressure of development and unprecedentedly fragile ecosystems. For this reason, environmental protection should be given priority. It is necessary to establish the scientific outlook on development, in a bid to build an environment-friendly society.

Vietnam has made remarkable achievement by maintaining a medium-and-high-speed GDP growth over the ten years. However, the environment indicators are revealing deteriorating environmental quality in Vietnam. Noticeably, industrial solid waste increased fastest to 7.5 million tons in 2010, nearly 12 times that of a decade ago. Meanwhile, the $CO_2$ emissions reached 43.95 million tons and organic water pollutant emissions 265,000 tons, increases of over three and five times, respectively. In short, Vietnam's economic growth is accompanied by environmental pollution, and industries develop at the expense of the environment.

Myanmar follows the same way to industrialization as its counterpart Southeast Asian countries, i.e., obtaining foreign exchange by exporting primary commodities, including primary agricultural and fishery products and mineral products. The most typical industries are primary resource-intensive industries and export-oriented import substitution industries of primary products processing. The industrial sector grows fast, including mining, energy, construction, and manufacturing.

As the above four countries have representativeness in development, national conditions, and land area, they are given detailed analysis. Other countries in the region all have their respective features with regard to green development. Brunei has strong reliance on oil and gas resources and the region's second highest per capita GDP. The forest coverage is high, and the per capita forest area reaches 12,001.41 $m^2$ and per capita forest biomass, 126.40 t. Brunei seeks economic diversification with weak reliance on oil and gas resources by developing industry and manufacturing.

Malaysia, Thailand, and China are above the middle level. Malaysian per capita GDP exceeds 10,000 US dollars and per capita forest area, 6000 $m^2$. Thailand's per capita GDP is slightly lower than that of China. While the per capita share of forest is low, the farmland is rich per capita and presents high rice yield. Tourism maintains a great momentum in recent years. The local governments pay attention to conservation of the existing ecological resources in the pursuit of economic growth and further sustainable development. Indonesia, Vietnam, and Laos, in the early stage of industrialization, lag behind China and Thailand. In Indonesia, forests which are largely evergreen broad-leaved reach more than 6000 $m^2$ per capita. Laos is economically backward, but has abundant natural resources, with per capita forest area as large as 12,590.04 $m^2$ and per capita forest biomass up to 270.81 t. Cambodia, similar to Myanmar, has a low per capita GDP, but a large coverage of natural vegetation. The vegetation area per capita reaches 6000 $m^2$ and farmland per capita exceeds 3000 $m^2$. Myanmar is rich in forest resources, with forest biomass per capita up to 7 t. The annual output of grain and oil attains 503 kg.

## 3.4 Opportunities of Green Development

Green development is challenged by two relations, namely the relation between rapidly increasing aggregate economic output and aggregate emissions and the relation between economic development and environmental capacity and quality in the fast process of urbanization. The evasion and mitigation of the two contradictions are important to green development.

### 3.4.1 Aggregate Economic Output and Aggregate Emissions

China and ASEAN are faced with the contradiction between rapid increase of economic output and control of aggregate emissions. Their economic output has maintained an upward trend over the years despite difference in growth rate. Because of cost and market factors, ASEAN has become the new target of capital investment in recent years, where a large number of international and Chinese companies set up factories instead of the original sites. Among them, Vietnam and Myanmar are most representative. In Vietnam, the total foreign direct investment of contract projects have increased, and a large amount of capital flows to the processing and manufacturing industries. By virtue of rich natural resources, Myanmar has also attracted a large number of investments from Asia, the UK and the USA, which markedly contribute to the local industrial upgrading and technological advancement. Myanmar's huge domestic consumer market and existing industrial chain and adjustable macroeconomic policies also drive the continued increase of the national gross product.

The growth of aggregate economic output is accompanied by the increase of aggregate pollutants. Currently, China ranks high in pollutant emissions, which is directly related to the execution of environmental protection policies. Air pollution has been very severe in Southeast Asia and under the existing model of development, will be further aggravated with the continued population growth in the process of industrialization. Water resources have also been destroyed in the context of urbanization, while resource consumption continues to increase in the ASEAN Member States. Another serious problem is rainforest shrinking at an alarming rate, which directly leads to increased GHG emissions and trans-boundary air pollution (e.g., haze). It is urgent to control the total pollutant emissions; otherwise, it will directly threat the health of residents and ultimately undermine economic development.

From the government point of view, the majority ASEAN Member States and China will have to deal with this contradiction in the first place of the environment-development relationship. On the one hand, impressive progress has been achieved in environmental protection, and on the one hand, livelihood related work has been complained about. The need to future increase the total economic output poses new requirements for industrialization and urbanization. As technology advances, the emission intensity, i.e., emissions per unit of economic output (or product yield), will be reduced, but the total emissions will remain high due to rapid economic growth. Extensive and unsustainable production and consumption patterns exacerbate the pressure on resources and the environment. The achievement of more demanding abatement targets will require greater efforts and huger investment. As time goes on, the base of emissions will increase, pushing up difficulty and cost of abatement. Hence, the pressure to achieve ambitious abatement targets without undermining economic growth has become the primary factor restricting the green development.

The contradiction between economic growth and environmental protection is quite typical in ASEAN and China. It is classified and described as follows.

Singapore is a densely populated and highly resource-poor island country. It continues land reclamation in order to meet the needs of expanding population and growing economy, given that the land area is not only small, but also from the perspective of geological structure, not suitable for cultivation, leading to reliance on imports of all agricultural products. Hence, the country has faced, since inception, the challenge to use limited resources to meet the growing demand in commercial, industrial, residential, and entertainment aspects.

China's economy shows excessive and increasing dependence on investment. Different from consumption and net exports, investment serves as a part of the total demand in the current period, and them a part of the total supply after put into use. The reality is that over-reliance on investment aggravates the serious overcapacity existing in a lot of industries and products. China has paid high resource and environmental costs for economic growth as the vast majority of products are resource-intensive. If not timely transformed, the production process will often cause serious environmental pollution and ecological damage.

In Vietnam, the worst environmental consequence of change in land-use patterns is increased extent and frequency of floods. Indeed, Vietnam is prone to floods, but the natural causes (monsoon rains and typhoons often affect the northern and central parts of the country) behind such natural disasters can be multiplied by improper human behavior or moderated by good human behavior. For example, floods in the Red River Delta and Mekong River Delta are directly related to the massive deforestation and serious soil erosion, and the situation continues to deteriorate. The statistics about large floods in Hanoi in the decade have proven that the increased flood levels can be attributed to destruction of forests with continued reclamation of woodland and wetland to develop agriculture and aquaculture.

Myanmar's economy mainly relies on agriculture which absorbs 60% of the labor force. From an environmental point of view, the large-scale development of agriculture will entail a substantial use of land, water, and other natural resources. Without appropriate management, soil and water resources are harmed with the extensive use of pesticides, and forest resources suffer severe damage in the process of industrialization and urbanization. In Myanmar, the forest coverage declined from 60% in 1990 to 50% in 2002 and further to 49% in 2005, indicating serious forest degradation. Due to resource exports, a large number of species are threatened with extinction. For example, Burmese padauk and Burmese tulipwood will soon be completely cut off according to the export rate in 2013. Myanmar's industrial enterprises, mostly of small scale and limited pollution treatment capacity, and foreign-funded enterprises, such as those in the jade industry, have also caused serious damage to the surrounding environment.

### 3.4.2 Relationship Between Economic Development and Urban Environment

Urbanization is accompanying the industrial revolution, and cities become the main place for social activities. Given certain socioeconomic and cultural levels, the environmental capacity of cities is subject to many restrictions, including land, water, and energy productivity, as well as science and technology development. In the process of rapid urbanization, environmental capacity and environmental quality are difficult to match the pace of economic development.

Urbanization remains mainstream in China and ASEAN in the past years and coming decade and is generally reflected in the priority development of large cities. In China, urban diseases are prominent in the current urbanization model, and the urbanization rate rose to 54% in 2014. In ASEAN, the urbanization rate also shows an upward trend, basically reaching 100% in Brunei, Malaysia, and Singapore, 40–60% in Indonesia and the Philippines, and 20–40% in other countries. The national differences in the level of urbanization are closely related to geographical location, historical background, economic base, and political regime, as well as policy implementation.

In China, the most prominent is social issues arising from universal low-density urban expansion in planning and construction, including traffic congestion, environmental pollution, functional chaos and population disorder. The widespread large-scale urban demolition and construction also lead to a serious waste of a lot of natural resources such as water and coal, direct pollution such as construction waste and air dust, and indirect pollution such as steel and cement. The entailed large demand of coal and oil gives rise to a lot of $CO_2$ emissions. In most ASEAN Member States, the population mainly concentrates in the capital city, such as Jakarta in Indonesia, which implies excessive concentration of urban population and economic activities. Excessive urbanization is one of the most serious urban problems and exerts a direct impact on the sustainable development of national economies. It will become grimmer with the rise of urbanization rate and population size. The consequences will include industrial backwardness, inadequate oxygen capacity, poor management, and faster population growth beyond the tolerance level of economic, environmental, and public facilities. More specifically, these problems can be viewed as contradictions between high pollution load and limited environmental capacity and between ambitious environmental protection targets and backward social development. They show the following characteristics in typical countries:

Singapore has created a beautiful and pleasant environment through the Clean and Green Singapore (CGS) campaign, at the expense of native terrestrial and marine habitats, ecological health, and indigenous organisms. This reflects the "ecopragmatist" idea of the Government which puts more emphasis on practical value than ecological value. "Nature areas" identified in the concept plan have no legal status and are vulnerable to erosion of other "strategic" and "national" benefits. A pragmatic approach is necessary for nature conservation due to limited land in the country.

Chinese per capita possession of primary resources is significantly lower than the world average, the average of developed countries and also average of middle-income countries and low-income countries. What's worse, ecological destruction is increasing with accelerated pace of industrialization and urbanization. The model of development that features high growth accompanied by environmental degradation will not sustain long-term high growth, but make more than one billion face the threat of deterioration of the living environment. China is currently the world's largest $CO_2$ emitter with considerable $CH_4$ and $N_xO$ emissions and therefore affects the global atmospheric environment. As a developing country, it does not assume obligations for reducing emissions under the Kyoto Protocol (under UNFCCC), but in the post-Kyoto Protocol era, China is likely to bear increasing international pressure, which, if not properly addressed, could affect its international image and status.

SMEs in Vietnam, particularly those specifying in paper making, industrial chemicals, oil refining, and steel production, generally lack technologies and capital to reduce noise, waste water, and waste gas. A variety of toxic waste emissions also increase with the expansion of the scale of these enterprises. Although Vietnam National Environment Agency (NEA) has developed standards for industrial waste gas or waste water, the vast majority of factories still directly discharge untreated waste into rivers, waterways, or ponds. For example, at least 20% of industrial solid waste in Hanoi is poured directly into dumps with no or only limited protective facilities. Water gas in Vietnam mainly includes $SO_2$ and $NO_x$ from thermal power stations and PM from coal mines and cement plants, but the emissions from non-key sources and small businesses increase substantially with the change in the economic structure brought by the industrial innovation. For example, the concentrations of CO, $CO_2$, dust particles, and lead in the atmosphere in major cities rise to worrying levels due to the extensive use of leaded gasoline in means of transport. In order to control pollution, NEA has taken such measures as shutting down Hai Phong Cement Plant and changing the way of waste emissions of gourmet powder processing plants in Ho Chi Minh City. However, the pollution index has not been reduced due to the absence of proper systemic management and lack of available resources.

Myanmar is rather special. It is the least economically developed with slow industrialization because of the political factor. Environmental governance seriously separates from and lags behind social development. With the end of closure and economic stagnation, many industrial areas were built after 1980, pushing up the level of industrialization, especially in the field of oil and natural gas. Despite the low level of industrialization overall, the resources, low costs and large market of Myanmar lure foreign investors. In view of relatively scarce electric power and increasing energy demand with rapid economic growth, oil and gas, and minerals that meet the needs of industrial development will serve as an important support for Myanmar's economic development.

### *3.4.3 China-ASEAN Regional Industrial Transfer*

The secondary and tertiary of ASEAN Member States are rather under-developed. Agriculture still takes the dominating role in their GDP with very low share of industry (manufacturing industry in particular) and service industry. In terms of the changes of their industrial structure, the share of agriculture decreased from 33.5% in 1970 to 12.99% in 2010 in the total GDP of ASEAN, but that of industry and service industry increased from 24.37 and 42.58 to 41.63 and 45.38% respectively. The change for industry is quite obvious with only a slight increase for service industry [3].

In recent years, China's labor and resource cost increased, but ASEAN has very rich labor resources, convenient transportation, government policy support, vast market and cheap land. Therefore, there emerged a trend of transferring the at least labor-intensive industrial chain from China to ASEAN region. This is the third round of industrial transfer following the transfer from Japan to four little tigers in Asia and the developed countries in Europe and America to China in late 1970s.

It is specifically demonstrated in the following levels: (a) gradient transfer and structural adjustment. China and ASEAN need a win-win out of the industrial harmonization. On the one hand, within China, there is a gradient transfer of the industry from east to the west; on the other hand, some industries are transferring to ASEAN Member States. (b) During the industrial transfer for professional division of labor, it is necessary to have an overall consideration of the industries and resources and allocate the resources in a rational way to avoid vicious competition of homogenized products and industries, mismatched development and locking the technology at the mid and low end. (c) There needs the specific carrier for the industrial transfer. Currently, the benchmarking of different industrial parks for industrial transfer could be a good way out.

Chinese government offers guidance for domestic companies to invest overseas in its industrial upgrading process, but the current support is not strong enough, and its policies are not favorable. External competition is another important factor affecting the transfer, in particular, the pressure from Europe, the USA and Japan whose multinational companies basically monopolized the economy of Southeast Asia and hope for more benefits out of their layout adjustment, which as a result posed great pressure on China. In fact, the takeover regions welcome more the green industry with less environment pollution but good economic benefits as they are more experienced and paying greater attention to sustainable development. The industrial transfer between different countries can serve as a big push for their economic development, and correct policies can also facilitate trade and investment growth and speed up the role of industrial restructuring and optimization. As a result of lowered competitiveness of Chinese products due to the appreciation of RMB and higher cost due to the constant rise of the price of oil, grain and raw materials on the international market, there is greater difficulty for the industrial transfer between China and ASEAN.

In recent years, China is investing more to ASEAN region, and now the investment targets gradually shift to the manufacturing, energy, and finance sectors. ASEAN is also the major region for China to attract foreign investment, largely from Singapore and Malaysia. The cooperation between China and ASEAN Member States is conducive to reducing their dependence on the developed countries in the west, helping them to realize industrial upgrade with their own resources and promoting the sustainable development of both.

In the process of industrial chain transfer, there are cooperation opportunities for China and ASEAN. For example, ASEAN needs massive infrastructure development. China has accumulated rich experience during its urbanization process which can be shared with ASEAN Member States. Though ASEAN unveiled many preferential policies to attract foreign investment, due to poor coordination among the member states, its overall competitiveness is undermined.

### *3.4.4 Global Integration and the Belt and Road Initiative*

In the context of globalization and regional economic integration, China and the ASEAN Member States are faced tough challenges in economic and social development, brought by various uncertainties at different stages of urbanization and industrialization. The common task of economic development and environmental protection opens up the possibility for cooperation on green development in the future. In this new mode of development, the two sides implement the sustainable development strategy to address the increasingly serious environmental pollution and ecological destruction and overcome the current limitations on environmental capacity and energy resources. Global cooperation on climate change toward sustainable development also provides an important opportunity for China and the ASEAN Member States to make a green low-carbon economic and social transition. Due to the low level of development, the advantage of late development can be better reflected in the region. Moreover, the huge market with regional integration provides favorable conditions for the rapid formation of the green industry chain. In inefficient traditional sectors, there is a lot of room to improve energy efficiency. Relying on China's powerful manufacturing sector and ASEAN's rich human capital, the region can serve as an attractive global center for green technology commercialization that creates opportunities for global green development. It is an important time to guide the new economic and technological change, in which advanced energy technologies and environmental protection technologies will become the focus of international technology competition and give rise to new economic growth points, new markets and new jobs.

The Belt and Road (B&R) Initiative launched by China enables all-round opening up and links the country to the Asia-Pacific Economic Circle. Given highly complementary economic and investment structures of neighboring countries, the initiative makes it geographically and strategically possible for industrial transformation and upgrading in the eastern coastal areas, and also cooperation with the ASEAN Member States. Further, the dividends brought by industrial upgrading and industrial transfer include investment opportunities, regional innovations owing to huge capital flow, and trans-regional cooperation on technologies, finance, and environmental protection. In this respect, cities will be the main space to accommodate future activities, and the pace of economic development will determine the level of living standards in the future.

China has been active in cooperation with ASEAN to promote green development in the region. Taking advantage of population and land area, China is vigorously opening up and strengthening environmental technology exchanges with neighboring ASEAN countries. By improving environmental standards, the country is actively involved in international cooperation on transfer of eco-friendly technologies, so that China and the ASEAN Member States can obtain more international capital and financial support. Moreover, China has developed cutting-edge green technologies in such areas as wind power, smart grid, low-carbon technologies, and high-speed railways, which lays a good foundation for future cooperation.

It is necessary to further strengthen cooperation within China-ASEAN FTA, such as speeding up investment in infrastructure for interconnection and further expanding cooperation in low-carbon energy. China's B&R initiative creates more opportunities for cooperation between China and ASEAN for multifaceted, multi-level, in-depth development. In 2000, the Chinese Government made the

FTA proposal and signed with ASEAN the Agreement on Trade in Goods and Services in 2004 and the Agreement on Investment in 2009. Following the establishment of China-ASEAN FTA in 2010, a number of policies and measures were rolled out. In 2013, China proposed to speed up China-ASEAN infrastructure development to increase interconnectivity and formally launched the B&R initiative. 2014 ushered in the second decade of the China-ASEAN strategic partnership, and 2015 witnessed the realization of economic integration in ASEAN. The rise of a more integrated market and production base enables more free capital flow within the region and makes ASEAN an integrated regional economic group. The FTA is expected to serve as a powerful engine of economic growth in the region, and the increased investment in trade will drive regional economic growth. China and ASEAN can conduct more cooperation in transportation, agriculture, forestry, and resources development, to learn from each other the experience of sustainable development.

The cooperation on green development between the two parties is influenced by a number of factors. With approximation of resource endowments, China and the ASEAN Member States face competition in international investment as both encourage exports to drive national development. The concerns from the ASEAN Member States about the loss of share of labor-intensive industries in the national market exert a direct effect on bilateral cooperation and undermine green development. Another unfavorable factor is China's unbalanced trade with ASEAN Member States, in which the trade volumes with Singapore, Malaysia, Thailand, Indonesia, the Philippines, and Vietnam take up over 95%, and with such countries as Brunei and Laos, less than 5%. In addition, resource issues and cultural differences will also affect the future green cooperation. It should be noted that the B&R initiative will reshape the existing cooperation framework by extending economic cooperation to cultural fields, which is deemed as a positive interaction on green economy.

In the future, China's economic restructuring will be driven by expanding domestic consumption, reducing domestic investment in infrastructure, developing technology-intensive industries and increasing the proportion of the service industry. The Internet Plus will impose a direct and significant impact on the entire production system by greatly improving the demand patterns of the consumption side. With the implementation of the B&R initiative, effective mechanisms will be established for China's eco-environmental cooperation with ASEAN, in fields of transportation, industries, energy and resource utilization, mineral development, logistics, and finance. The two sides can promote win-win economic development by attracting green investments, guiding foreign capital flows to new energy and other green industries, increasing green investment to ASEAN, establishing biomass energy industry bases, as well as strengthening technology transfer and promotion.

## References

1. Zhou, G., et. al.: Regional Environmental Protection International Cooperation Strategy and Policy—Observation and Research on Asia-Pacific Environment. China Environment Press, Beijing (2014)
2. UNDP: Human Development Report 2014—Sustaining Human Progress: Reducing Vulnerabilities and Building Resilience. http://hdr.undp.org/en/content/human-development-report-2014 (2014)
3. Peng, B., Liu, X., Yang, Z.: Status of Resources and Environment and Potential of Cooperation in ASEAN. Social Sciences Academic Press, Beijing (2013)

# Chapter 4
# Policy Measures for Regional Green Development

**Abstract** This chapter reviews the environmental policies and mechanisms in China and ASEAN and identifies most commonly used policy measures taken by the two sides for green development. China and ASEAN's environment governance capacity has been improved constantly as a result of socioeconomic development and international environmental actions, as well as in response to the requirements for environmental protection. They established national agencies to address environmental issues and at the same time made good use of regional cooperation mechanism. With future economic and social development, the emergence of new environmental issues, and the increased public awareness for environment protection, greater potential will be released for further development of national environmental governance system and regional environmental cooperation mechanisms for China and ASEAN Member States. Multilateral environmental agreements are the foundation of the solutions to global and regional issues. China and ASEAN are active in joining many multilateral regional and international environmental agreements pertaining to climate change, biodiversity, water pollution, air pollution, chemicals and wastes, coastal and marine resources, etc. China and ASEAN Member States developed and implemented many country-specific environment laws, action plans, and policy measures. If environmental policies of one country can generate promising economic, social, and environmental benefits, they can serve as reference for others. This chapter then evaluates these policy measures and proves that policy interventions did play a significant role in addressing environmental issues.

## 4.1 Introduction

This chapter intends to present an overview of environmental policies in China and ASEAN and evaluate policy measures taken by China and ASEAN to promote green development, particularly identified widely used policies, and shortlist the most promising policy measures suitable for further analysis according to policy effects. Policy recommendations in favor of regional sustainable development will



China-ASEAN Environmental Cooperation Center, *China-ASEAN Environment Outlook 1 (CAEO-1)*, DOI 10.1007/978-981-10-6211-7_4

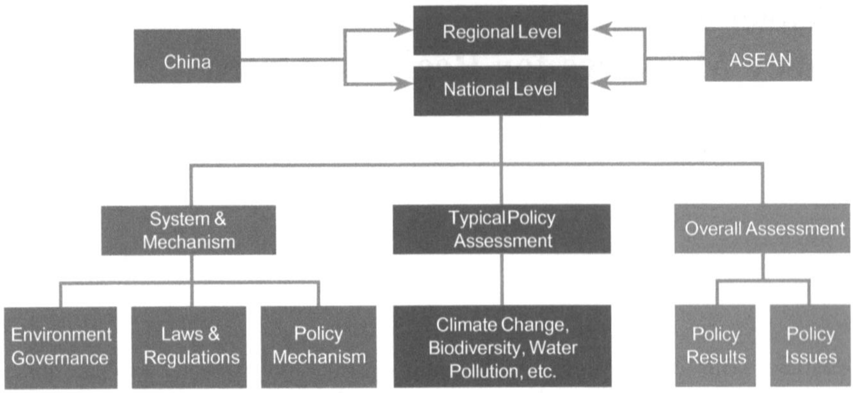


**Fig. 4.1** Framework of policy measures for regional green development

be provided, considering the benefits and limitations of policy combinations in different fields.

The work flow of this chapter is shown in Fig. 4.1. First, research reports, statistical data, and other information are collected, followed by the analysis of the policy drafting systems and mechanisms for green development in China and ASEAN, including environmental treatment system, laws and regulations, and policy mechanism, etc. Third, typical policy measures are analyzed taking into account major global and regional environmental issues like climate change, biodiversity, and freshwater resources. And in the end, the overall judgment of the policy measures of the two sides is developed.

## 4.2 Overview

### *4.2.1 Environmental Governance System*

#### 4.2.1.1 Regional Environmental Governance System

ASEAN has established itself as an institutionalized regional cooperative organization. As early as 1970s, ASEAN incorporated environment consideration into its integration process and shaped the multi-level cooperation framework for environment issues. Currently, there is a fairly full-fledged mechanism framework for ASEAN environment cooperation, including ASEAN Summit, ASEAN Ministerial Meeting on the Environment, ASEAN Senior Officials Meetings, and ASEAN Secretariat.

China-ASEAN cooperative mechanism for environment is established gradually. With the economic and social development of China and ASEAN, environment protection was put on agenda. In 2003, the two sides signed Joint Declaration on ASEAN-China Strategic Partnership for Peace and Prosperity which further highlighted the cooperation on environment protection. The ASEAN-China Strategy on

Environmental Protection Cooperation 2009–2015 adopted in 2009 listed environmentally sound technology, environmental labeling, and cleaner production as well as environmental goods and service as the priority areas for cooperation. In 2011, China and ASEAN drafted ASEAN-China Environmental Cooperation Action Plan (2011–2013) which proposed to conduct the exchanges on environmental industry and technologies and establish ASEAN-China environmental industry cooperation network. In order to further strengthen the dialogue and cooperation between the two sides on environment protection, China and ASEAN jointly launched an annual ASEAN-China Environmental Cooperation Forum in 2011 in Nanning, Guangxi Zhuang Autonomous Region. The forum has become an important platform to conduct dialogue, facilitate exchanges, and promote cooperation on environment between China and ASEAN.

Some ASEAN Member States conducted collective regional environment cooperation with China. In 1992, Asian Development Bank (ADB) held the first Greater Mekong Subregion (GMS) Ministerial Conference in its headquarters in Manila, the Philippines, and kicked off the GMS mechanism officially. Currently, GMS cooperation covers China (Yunnan province and Guangxi Zhuang Autonomous Region), Cambodia, Laos, Myanmar, Thailand, and Vietnam. In the past 20 years, the GMS cooperation has been progressing smoothly with cooperative projects conducted in nine priority areas including environment. It has been active in providing financial support and technical assistance for member countries and carried out many initiatives including Core Environment Program and Biodiversity Conservation Corridors Initiative in the Greater Mekong Subregion.

#### 4.2.1.2 Environment Governance Systems of China and ASEAN Member States

China and the ASEAN Member States have put in place a relatively complete system of environmental governance, and even the late-start countries have established agencies specially to deal with environmental issues. The environmental governance system continues to improve with the emphasis on environmental protection.

State-level environmental management departments have been established in China and the ASEAN Member States. With the growing impact of environment on health and economy, environment has been a worldwide concern, and the establishment of environmental protection agency itself reflects the government's commitment to tackling environmental problems. In China and most ASEAN Member States, independent environmental protection agencies were established in subsidiary departments in the 1980s by integrating new duties and responsibilities with those transferred from existing government agencies. To date, the departments in charge of national environmental affairs are in place, though named differently. As shown in Table 4.1, Indonesia was first to establish the ministry of environmental protection, specifically in 1973, and Myanmar and Laos started late and completed the agencies in 2012.

**Table 4.1** Inception of environmental management departments in China and ASEAN

| Country | Department | Time of establishment |
|---|---|---|
| China | Ministry of Environmental Protection | 2008 |
| Brunei | Environment, Parks and Recreation Department | 1993 |
| Cambodia | Ministry of Environment | 1996 |
| Indonesia | Ministry of Environment | 1973 |
| Laos | Ministry of Natural Resources and Environment | 2012 |
| Malaysia | Ministry of Natural Resources and Environment | 2004 |
| Myanmar | Ministry of Environmental Protection and Forestry | 2012 |
| The Philippines | Ministry of Natural Resources and Environment | 1987 |
| Singapore | Ministry of Environment and Water Resources | 2004 |
| Thailand | Ministry of Natural Resources and Environment | 2002 |
| Vietnam | Ministry of Natural Resources and Environment | 2002 |

It should be noted that the establishment of environmental agencies, departments, or ministries in these countries is closely linked to the global developments in environment policies and governance which came out as a result of global conferences on environment and development, as well as their respective environmental problems arising from development.

**Global and regional developments in environment and green development policies and governance**

Initial stage (1972–1992): In June 1972, the Declaration on the Human Environment was issued and the Action Plan for the Human Environment adopted at the first United Nations Conference on the Human Environment held in Stockholm, Sweden, raising public environmental awareness and triggering the surge of worldwide environmental movement. It was the time when China and some ASEAN Member States kicked off environmental protection: the State Council Leading Group for Environmental Protection of China set up in 1973, the Singapore Ministry of the Environment established in September 1972 (renamed as the Ministry of the Environment and Water Resources in 2004), and the Indonesia Ministry of Environmental Protection established in 1973.

Development stage (1992–2002): In June 1992, the United Nations Conference on Environment and Development took place in Rio de Janeiro, Brazil, proposing a sustainable development strategy that for the first time combines economic development with environmental protection, and stressing strengthened international cooperation to jointly cope with global environmental issues. At this stage, more countries started to pay attention to environmental issues. In 1987, the Department of Environment and Natural Resources was founded in the Philippines, in 1988 the State Environmental Protection Administration in China, and in 1993 Environmental Protection Agency in Brunei. Thailand included environment as responsibility of the Ministry of Science, Technology and Environment (formerly known as Ministry of Natural Resources and Environment) in 1992.

Adjustment stage (2002–2008): In August 2002, the World Summit on Sustainable Development held in Johannesburg, South Africa, called for economic growth and social progress in harmony with environmental protection and ecological balance. The status of environmental protection departments was further enhanced with the adjustment to their functions. In 2002, the Congress of Vietnam passed a resolution on establishing the Ministry of Natural Resources and Environment, and the Thai Government re-organized the Ministry of Science, Technology and Environment into Ministry of Natural Resources and Environment as a special agency for resource management and environmental protection. In 2008, China upgraded the State Environmental Protection Administration to Ministry of Environmental Protection (MEP) as an integral department of the State Council.

Perfection stage (2008–2012): In June 2012, the United Nations Conference on Sustainable Development held in Rio de Janeiro considered green economy as an important tool available for achieving sustainable development. In the same year, Myanmar formed the Ministry of Environmental Protection and Forestry and Laos set up the Ministry of Natural Resources and Environment. To 2012, environmental management departments were put in place in China and ASEAN Member States, though at different times.

In addition to these four environmental conferences, environmental problems arising from development also pushed forward the establishment of environmental management agencies. For example, Malaysia set up the National Forestry Council in 1971 for the purpose of managing forests in the face of serious deforestation. Singapore's National Water Agency was created in 1979 to address severe water shortage. Brunei found a specialized agency for smog pollution control, as a result of haze pollution caused by Indonesia's forest fires in 1997. In Cambodia, natural reserves have been created early to address biodiversity loss. Environmental issues attract the attention of governments and stimulate the formation of environmental management departments to meet the growing demand for specialized environmental management.

At the regional level, an ASEAN Resolution on Environment and Development was issued in February 1992 in Singapore to intensify cooperation in environmental management and protection in their common pursuit of sustainable development. In this resolution, ASEAN Member States agreed, among others, to: (a) introduce policy measures and promote institutional development that will encourage the integration of environmental factors in all developmental processes; (b) work closely on the inter-related issues of environment and development; (c) cooperate in setting basic environmental quality standards and regulations at national level, work toward harmonized environmental quality standards in the region, and adopt long-term quantitative goals relating to ambient air quality and river water quality; and (d) harmonize policy directions and step up operational and technical cooperation on environmental matters such as transboundary air water pollution, natural disaster, forest fires, oil spills, and the transboundary movements and disposal of toxic chemicals and hazardous wastes, and undertake joint actions to address the anti-tropical timber campaign.

As can be seen from the course of development, the inception of environmental management departments in China and ASEAN Member States is driven mainly by two factors: international concerns about environmental issues and environmental problems arising from national developments. China and ASEAN Member States have placed enhancing emphasis on environmental issues and improvement of environmental management system.

**Institutional characteristics**

There is no universal model for the setup of the environmental protection agencies. However, some common features can be summarized as follows.

Environmental protection departments take charge of environmental protection with support from related departments. In China, in addition to MEP, other ministries under the central government also assume responsibility for environmental management. For example, the Ministry of Land and Resources is responsible for the protection of land resources, Ministry of Water Resources for the protection of water resources, Ministry of Agriculture for the protection of agricultural environment and aquatic organisms, and State Forestry Administration for the protection of forest resources and wildlife. Similar to China, the most ASEAN Member States also assign environmental responsibility to other related departments, in order to create synergy. In Indonesia, Ministry of Environment assumes responsibility for administration of the environment and natural resources, while Ministry of Marine Affairs and Fisheries is responsible for improving the quality of marine and coastal environment, islands, and freshwaters, Ministry of Forestry for the protection of forest resources, and Water Authority for safeguarding sustainable development of water resources. In Thailand, similarly, the responsibility of environmental protection is shared by the Ministry of Natural Resources and Environment and other relevant ministries. Specifically, the Ministry of Agriculture and Cooperatives manages key natural resources and flora and fauna habitats; Ministry of Industry is responsible for plant pollution control and management; and Ministry of Transport and Communications implements vehicle emission control programs.[1]

The functions of environmental protection agencies are clearly defined, which are roughly the same with slight differences among countries. Though expressed in different ways, the responsibilities of environmental protection agencies basically cover the following aspects: development of environment-related laws, regulations and policies, supervision and management of environmental pollution control, environmental data collection and management, environmental monitoring and information disclosure, international exchange and cooperation, and environment-related publicity and education. Apart from the similarities, there are differences in specific responsibilities. For example, China provides for supervision and management of nuclear and radiation safety; Cambodia defines the administration of

[1]Asian Development Bank (2006): Country Synthesis Report on Urban Air Quality Development.

national nature reserves; Malaysia includes management training and human resource development into the responsibility for ministry of environmental protection; and Laos rewards environmentally well-performing individuals or organizations.

The divisions of environmental protection agencies vary among countries, covering different ranges of aspects. The organizational structure of China's MEP is similar to that of Cambodia and Brunei. In the Philippines, Malaysia, and Thailand, a wider range of responsibilities are covered and their organizational frameworks are similar, and closer to the so-called super-ministry. In these countries, in addition to divisions for general affairs, the specialized divisions are set according to environmental factors. Mines, forests, and other natural resources fall into responsibility of the environmental authorities.

A top-down hierarchical management system is established. In China, MEP at the central level is supported by environmental protection agencies at the local level, including bureaus in provinces, autonomous regions, and municipalities and offices or specialized agencies of cities, counties, and county-level cities. Local agencies take charge of environmental protection within their jurisdiction, under the leadership of local governments and higher-level environmental agencies. In Cambodia, the environmental management system integrates central and provincial levels, in which the Ministry of Environment under the central government represents the highest level environmental administration. Local agencies include environmental offices of provinces and cities and implementing and/or supervision agencies in districts and communities.

Dedicated agencies and coordinating bodies are established for transboundary environmental management. To address environmental problems free from the restriction of administrative boundaries, such as haze pollution, water pollution, and biodiversity loss, appropriate transboundary or inter-regional environmental management agencies are needed for better coordination. From 2006 onwards, China has set up six environmental supervision centers, respectively, overseeing the south, southwest, northeast, northwest, east, and north areas. The responsibilities of these agencies are to coordinate and handle inter-provincial or inter-regional major environmental disputes and to accept, coordinate, and handle visits and complaints on inter-provincial or inter-regional environmental pollution and ecological damage, which provide strong support for regional environmental regulation. In addition, there are cross-regional river basin commissions established for major rivers. For example, China's Yellow River Conservancy Commission fulfills management duties in the Yellow River Basin, Xinjiang, and Qinghai, and assumes responsibility for rational development, utilization, and protection of water resources in the basin. ASEAN also established several mechanisms for transboundary cooperation, e.g., ASEAN Agreement on Transboundary Haze, ASEAN Center for Biodiversity (ACB), the Greater Mekong Subregion (GMS) framework, and the Heart of Borneo. The Mekong River Commission, as an inter-governmental agency for cooperation among Thailand, Laos, Cambodia, and Vietnam, stresses the sustainable development of the Mekong River Basin and facilitates management of river resources, fisheries, and environmental protection in the basin.

In short, in response to international environmental issues and domestic environmental problems, China and the ASEAN Member States have put in place relatively complete systems of environmental governance and established institutional frameworks that are specific to their national characteristics and conditions. Moreover, the continuous improvement of the environmental system implies that countries take environmental issues more seriously.

## 4.2.2 Environmental Legal and Policy Framework

### 4.2.2.1 Environmental Laws

The environmental legal system of China and the ASEAN Member States generally includes the legal basis at two levels, namely basic law at the national level and laws specific to elements of the environment.

Basic law: China and ASEAN Member States have drafted a series of environmental laws, and almost all developed the basic law. As early as in 1968, Singapore enacted a dedicated environmental protection law, and Vietnam, a late-starter of environmental protection, adopted its Environmental Protection Act in 1999. By far, there have not yet been integrated environmental laws in Brunei and Myanmar as the former puts in place special laws on environmental factors and the latter incorporates environmental provisions into other laws and regulations (see Table 4.2).

Separate laws: The separate laws on environmental protection in China and ASEAN can be roughly categorized into pollution prevention and control and environmental resources. In the first category, examples are the Law of China on Air Pollution Prevention and Control, Water Quality Standard and Water Pollution Control Act of Thailand, and Ecological Solid Waste Management Act of the Philippines. In the second category, examples include various water resources acts, mining acts, and wildlife conservation acts in these countries. These different separate environmental laws basically tackle the major environmental problems faced by China and ASEAN Member States, such as air pollution, biodiversity, and water environmental management (see Table 4.3).

### 4.2.2.2 Environmental Policies

Appropriate environmental policies are also rolled out to cope with emerging environmental issues. According to the World Bank criteria, they can be classified into command-and-control policy, environment-economic policy, and voluntary policy.

The command-and-control policy refers to traditional administration that directly regulates emissions of specific pollutants by managing production processes or product use or regulates the behavior of environment pollution within specific time and region. The regulation of production processes and products is achieved mainly through adoption of technical standards and deadlines for governance. Emphasizing

**Table 4.2** Promulgation of environmental laws in China and ASEAN

| Country | Basic law | Promulgation and amendment |
|---|---|---|
| China | Environmental Protection Law | Adopted at the 11th Session of the 7th NPC Standing Committee on December 26, 1989 and amended at the 8th Session of the 12th NPC Standing Committee on April 24, 2014 |
| Brunei | None | Absence of a comprehensive environmental code |
| Cambodia | Environmental Protection and Natural Resources Management Act | First law on environmental protection passed by the National Assembly of Cambodia in 1996 |
| Indonesia | Environmental Management Act | Enacted in 1997, replacing the Environmental Basic Law 1982 |
| Laos | Environmental Protection Act | Formally implemented April 26, 1999 |
| Malaysia | Environmental Quality Act | Promulgated by the Malaysian Government in 1974 |
| Myanmar | None | Myanmar currently has not developed detailed environmental management support legal or comprehensive plan of action on environmental regulations are contained in a series of laws and regulations among |
| The Philippines | Philippine Environmental Policy<br>Philippine Environmental Code | Formally promulgated in 1977 |
| Singapore | Environmental Public Health Act | Enacted in 1968 and amended several times afterward |
| Thailand | Governance Promotion and Protection Act | Formulated and implemented in 1975 and amended in 1978, 1979, and 1992 respectively |
| Vietnam | Environmental Protection Act | Enacted in 1993 and put in force on January 1, 1994. The amendment passed by the Congress on November 29, 2005 includes relatively large adjustment and refinement of the act |

end-of-pipe pollution control, the policy prohibits or restricts emissions of specific pollutants and defines the spatial and temporal scope of polluting activities. It mainly encompasses emissions standards and permits, pollution control regulations, and environmental supervision and enforcement regulations.

The environment-economic policy guides economic shareholders to acts favorable for environment by way of cost-effectiveness. Typically, it involves payment or money transfer or creation of new markets. According to the principles of environmental economics, a variety of economic levers, including pricing, taxation, credit, investment, microeconomic stimulus, and macroeconomic regulation, are available to adjust or influence environmental behavior of stakeholders. The policy instruments include environmental taxes and charges, environmental fiscals, and green trade.

**Table 4.3** Separate environmental laws in China and ASEAN

| Country | Main environmental single laws |
|---|---|
| China | Air Pollution Prevention and Control Law, Water Pollution Prevention and Control Law, Solid Waste Pollution Prevention and Control Law, Wildlife Protection Law, Environmental Noise Pollution Prevention and Control Law, Environmental Impact Assessment Law |
| Brunei | Oil and Mining Act, Forest Act, Water Supply Act, Urban and Rural Planning Act, Wildlife Act, Land Act |
| Cambodia | Solid Waste Management Act, Water Resource Management Act, Air Pollution and Noise Management Act |
| Indonesia | Forest Management Basic Law, Mining Act |
| Laos | Water and Water Resources Law, Land Law, Mining Law |
| Malaysia | Land Protection Act, Water Act, Fisheries Act, Wildlife Protection Act, Solid Waste and Public Cleansing Management Act |
| Myanmar | Wildlife and Natural Resources Protection Act, Forest Law, Groundwater Act, Marine Fisheries Law |
| The Philippines | Clean Air Act, Ecological Solid Waste Management Act, Wildlife Protection Act, Water Purification Act |
| Singapore | Wild Animals and Birds Act, Hazardous Waste Act, Endangered Species Act |
| Thailand | Atmospheric Standards and Air Pollution Control Act, Water Quality Standard and Water Pollution Control Act, Chemicals, Hazardous Waste, and Hazardous Substance Act, Forests Act, Wildlife Act, Mining Act, Environmental Analysis and Assessment Act |
| Vietnam | Forest Protection and Development Act, Land Act, Oil Act, Water Act, Mining Act |

The voluntary policy refers to environmental charters, code of environmental conduct, and environmental management standards initiated voluntarily by international organizations and industry associations to promote improvement of the environmental behavior of polluting enterprises. Not bound by mandatory laws and regulations, this policy serves as standards and rules for voluntary companies to improve their environmental performance. It affects corporate environmental behavior by encouraging public participation and social supervision on the grounds of the public right to know the environment so that enterprises are forced to standardize their environmental behavior and cut more pollutant emissions. The policy covers information disclosure, environmental labeling certification, and International Organization for Standardization (ISO) Management System certification.

The command-and-control policy currently serves as the leading environmental policy in China and ASEAN. The environmental policy in most countries mainly relies on a command-and-control approach, of which administrative order is the most common form of issuance. Currently, the command-and-control policy plays a dominant role, while the environment-economic policy is under progressive development. The voluntary environmental policy is mainly applied in the pilot and remains in the exploratory stage. The command-and-control policy is mainly embodied in pollution control standards, EIA policies, and others (see Table 4.4).

**Table 4.4** Typical command-and-control policy in China and ASEAN

| Country | Pollution control standards | EIA policies | Others |
|---|---|---|---|
| China | Air Pollutant Emissions Standards, Wastewater Discharge Standards, Cleaner Production Standards | Technical Guidelines for Environmental Impact Assessment, Planning, Environment Impact Assessment Regulations | Nature Reserves Regulations, Environmental Noise Pollution Regulations |
| Brunei | – | No clearly defined EIA system | "Lead-free" petrol policy |
| Cambodia | Motor vehicle emissions standards | Administrative regulations related to EIA procedures | Administrative regulations on solid waste management, water pollutant management, air pollutants and noise management |
| Indonesia | Noise Standards DME48/1996, Air Pollutant Standards DME15/1996, Water Quality Standards | No specific EIA law, but only relevant provisions included in Act 22/1999 and Regulations 25/2000 | Water Pollution Control Measures, Toxic and Hazardous Waste Regulations 18/1999 |
| Laos | – | Laos' EIA system to be standardized | – |
| Malaysia | Motor Vehicle Emissions Standards, | – | Environmental quality regulations, National Forestry Policy |
| Myanmar | Motor Vehicle Emissions Standards (the only pollution control standard) | Under preparation | Pollution Control and Cleaning Regulations, Public Health Act, National Forestry Policy |
| Thailand | – | No complete EIA system, but applicable strategic environmental assessment (SEA) principles and methods | – |
| Vietnam | Auto emissions standards | – | Biodiversity Action Plan, regulations on the control and management of hazardous waste |

*Note*—means no relevant policy established or identified in the country

China and ASEAN are experimenting on the environment-economic policy which is showing positive results. At present, a relatively efficient environment-economic policy is put in place only in some countries, and market-based instruments are yet to extend to all ASEAN Member States. The majority of ASEAN Member States have just started introducing the environment-economic policy.

The earliest environment-economic policy implementation introduced by China is the pollution charge system which reduces emissions of pollutants by charging emitters. It draws corporate attention to environmental protection and encourages technological transformation for comprehensive utilization of resources so that enterprises can play an active role in environmental protection while developing production. In addition, pollution charges provide strong support for the country's environmental protection by supplying a lot of funds (see Box 4.1).

**Box 4.1 China's pollution charge system**

China's pollution charge system formally took shape with the issuance of Interim Measures for Pollution Charges in July 1982 and implementing rules introduced in provinces, autonomous regions, and municipalities. In January 2003, the State Council promulgated the Regulations on Collection and Management of Pollution Charges and subsequently issued the Administrative Measures on Pollution Charges Standards and Administrative Measures on Collection and Use of Pollution Charges. After the implementation of the regulations, pollution charges grew from 6.74 to 19.8 billion yuan in 2013. In 2014, the pollution charges were increased dramatically in China. The national requirement is one time higher than the 2003 standard, and the local government can issue even higher pollution charges. For example, the charges in Beijing are 10 times higher than the national level in 2003. Higher pollution charges spurred the enterprises to better control pollution. Due to enterprises' better performance and growth slowdown, in 2015, the total national pollution charges decreased to 17.3 billion yuan which demonstrates an important role of pollution charges to pollution control.

- Legal basis for pollution charges: Air Pollution Prevention and Control Law, Environmental Protection Law, Water Pollution Prevention and Control Law, Solid Waste Pollution Prevention and Control Law, and Environmental Noise Pollution Prevention and Control Law.
- Targets: industrials and privately or individually owned business that directly discharge pollutants into the environment
- Use of pollution charges: subvention or discount loans for the following projects:
- Prevention and control of key source pollution
- Regional pollution prevention and control
- Development, demonstration, and application of new technologies and processes for pollution prevention and control
- Other pollution prevention and control projects under the State Council.

Source: Chinese Ministry of Environmental Protection Web site

Apart from the pollution charge system, desulfurization price subsidies for power plants and urban sewage treatment fees also make outstanding contributions to emissions reduction. For the control of $SO_2$ emissions, China provides a 0.015 yuan/(kW•h) tariff discount for coal-fired units and raises emissions standards to promote green power generation. This move has played a positive role in the installation and operation for desulfurization devices, which leads to a 3.40% decline of national $SO_2$ emissions in 2014 over the previous year. In addition, urban sewage treatment fees and improving charging criteria stimulate the construction and operation of sewage treatment plants, and as a result, the COD emissions were effectively reduced by 2.47% in 2014 from 2013 levels.

Singapore's water pricing policy effectively guides consumers to conserve water. While pricing reflects the commercial value and scarcity of water resources, water tax is levied to encourage the public to save water. The per capita daily water consumption in the country was cut to 165 L in 2003 and 156 L in 2008 from 172 L in 1995, according to statistics, and the government aims to, by 2030, reduce the level to 155 L.[2] In addition, subsidies are provided to encourage water-saving behavior. For example, projects that halve water consumption will be given a subsidy equal to half of the total investment in equipment.

Green procurement in Thailand gains a strong momentum in recent years. Under the Program of Thai Green Supply Chain Development, the green procurement Web site that defines "green procurement" and "green products" was created in August 2004. This Web site collects and publishes information related to "green products" and provides "best practice" guideline on green procurement procedures, which contributes to raising consumer awareness of green procurement and green products.

The environment-economic policy, though not yet prevalent, shows an obvious trend of development in ASEAN. For example, in 2005, the Vietnamese government began to impose mining environmental fees at a rate of 2000–50,000 VND/$m^3$ or 1.500–30,000 VND/t, covering ore mining, sand excavation, coal mining, and natural mineral spring water. Cambodia plans to establish the pollution charge system in an effort to control pollutant emissions by economic means.

The voluntary environmental policy begins to perform in China and some ASEAN Member States. The information disclosure system advances in recent years. With the assistance of the World Bank, China and Indonesia piloted the information disclosure system among enterprises in the 1990s. Since then, the system has been extended to provinces in China, and a series of relevant policies rolled out, including the Measures on Environmental Information Disclosure (Trial) and Measures on Self-Monitoring and Environmental Information Disclosure for Key Enterprises (Trial), in order to encourage enterprises to consciously protect the environment in the production process. In addition, China presses ahead with environmental labeling certification and ISO Environmental Management System certification. Starting in 1994, the environmental labeling work covers

[2]Source: Fourth ASEAN State of the Environment Report (SoER4), P34.

determination of product categories and development and promulgation of certification standards. At the end of 2014, totally 107 technical specifications for environmental labels on products were introduced, guiding and promoting energy conservation in related industries (see Box 4.2).

**Box 4.2 China's environmental management system certifications**

In May 1997, China Steering Committee for Environmental Management System Certification (CSCEC) was established with the approval of the General Office of the State Council. CSCEC provides guidance on implementation of the ISO environmental management system standards (ISO 14000) in China, of which ISO 14001 specifies environmental management system standards. In 2001, there were 1000 organizations certified under ISO 14001. According to the ISO survey, with a total of 91,590 certificates, China ranked first among large ISO 14001 certification issuers in 2012.[3]

In short, China and ASEAN have relatively complete legal and policy systems. Apart from the basic environmental law, a series of separate environmental laws are introduced to deal with specific problems, and meanwhile, appropriate policy measures play an important role in environmental protection. The current environmental policy mainly relies on regulation. Among them, the environment-economic policy and voluntary policy have been effective in some countries and attracted increasing attention of decision-makers, implying large room for development.

## 4.2.3 Environmental Planning and Action

### 4.2.3.1 Environmental Planning

Environmental planning laid out environmental protection targets and measures to guide people to protect the environment, and on this basis, the action plan specifies the program and actions. Whether for general or specific environmental issues, China and ASEAN have developed a series of environmental planning and action plans.

China's environmental planning, which can be dated back to the era of planned economy, refers to the five-year environmental plan renewed once every five years since the 1970s in the era of market economy. The 12th Five-Year Plan for Environmental Protection set down targets to be achieved by 2015: (a) drastic reduction in emissions of major pollutants; (b) effectively safeguarded environment

[3]The ISO Survey of Management System Standard Certifications—2012, http://www.iso.org/iso/iso_survey_executive-summary.pdf.

and greatly improved water quality of urban and rural drinking water sources; (c) effective control of heavy metal pollution and significant prevention and control of POPs, hazardous chemicals, and hazardous wastes; (d) improved levels of urban environmental infrastructure construction and operation; (e) reversion of ecological deterioration trend; (f) markedly improvement in nuclear and radiation safety and the regulation; and (g) perfection of environmental monitoring system. Box 4.3 lists the main content of China's 12th Five-Year Plan for Environmental Protection.

**Box 4.3 Main content of China's 12th Five-Year Plan for Environmental Protection**

- Reduce emissions of major pollutants

– Intensify structural adjustment
– Focus on the reduction of COD and ammonia nitrogen emissions
– Cut the intensity of $SO_2$ and $NO_x$ emissions.

- Effectively solve outstanding environmental problems

– Improve water environmental quality
– Implement integrated control of air pollutants
– Enhance soil environmental protection
– Beef up ecological protection and supervision.

- Strengthen environmental risk prevention in key areas

– Press ahead with process-wide environmental risk management
– Enhance nuclear and radiation safety management
– Contain the momentum of heavy metal pollution incidents
– Promote safety disposal of solid waste
– Perfect environmental risk prevention and control system for chemicals.

- Improve the basic public service system of environmental protection

– Promote the equalization of basic public services for environmental protection
– Elevate the level of environmental protection in rural areas
– Beef up the environmental supervision system.

- Carry out major environmental projects

– Optimize policy measures
– Implement the environmental responsibility system
– Improve the integrated decision-making mechanism
– Firm up the legal system
– Better environment-economic policies
– Consolidate scientific and technological support
– Develop environmental protection industries
– Boost investment

  - Provide strict law enforcement supervision
  - Mobilize the local people's governments
  - Promote departmental coordination in environmental protection
  - Actively guide public participation
  - Strengthen international environmental cooperation.

- Enhance organizational leadership and evaluation

  Source: China's 12th Five-Year Plan for Environmental Protection

In addition, a number of action plans are issued to address specific environmental issues. They include the China National Biodiversity Conservation Strategy and Action Plan (2011–2030) unveiled in 2010 to cope with new problems facing biodiversity conservation and the Action Plan for Air Pollution Prevention and Control and Action Plan for Water Pollution Prevention and Control released in 2013 and 2015, respectively, to improve air and water quality. Among the action plans under preparation, the Action Plan for Soil Pollution Prevention and Control was passed at the MEP Executive Meeting in March 2015 and is expected to publish in 2016.

Similarly, Brunei has developed the National Development Plan every five years from 1966 onwards, which fully reflects the government's emphasis on environmental protection. For example, the 5th National Development Plan included the policy of protecting tropical rain forests and biodiversity and the 6th and 7th provided specific action plans for the national environmental protection strategy. Indonesia has taken "national action to improve forest and land" since 2003, intending to build 300,000 ha forests along 29 rivers and in fact realizing 3 million ha of forests along 68 rivers during 2003–2007. Singapore defines "four national taps" (i.e., catchment water, water imported from Malaysia, NEWater, and desalinated water) and develops the long-term plan to, by 2060, triple NEWater supply to meet 50% of water demand and increase at least tenfold desalinated water supply to meet 30% of water demand.

In 1998, ASEAN adopted Hanoi Plan of Action (1999–2004), namely ASEAN Strategic Plan of Action on Environment (1999–2004). It set the overarching objective and specific plans for the future environment protection and sustainable development in ASEAN region. In 2002, the 7th Informal ASEAN Ministerial Meeting on the Environment established the 10 prioritized cooperative areas and agreed to have each country to take the lead for the cooperation in one area. At the ASEAN Environmental Ministerial Meeting in 2007, countries agreed to incorporate sustainable forest management, and sustainable management of natural parks and protection zones into the sustainable management of biodiversity. Currently, the environment cooperation in ASEAN is focusing on the 10 priority areas determined by ASEAN Socio-Cultural Community Blueprint (2009–2015) adopted in 2009, namely addressing global environmental issues, managing and preventing transboundary environmental pollution, environment education and public participation, environmentally sound technology (EST), urban environment management and governance, harmonizing environmental policies and databases, promoting the sustainable use of coastal and marine environment, promoting sustainable management of natural resources and biodiversity, sustainable freshwater resource management, responding to climate change, etc.

The environmental planning of China and ASEAN Member States have put forward specific and clear targets for the improvement of environmental quality, reflecting the determined environmental governance and laying the basis for environmental work and performance evaluation.

#### 4.2.3.2 Multilateral Agreements

Multilateral environmental agreement refers to agreements on certain environmental issues signed by and among two or more countries and generally led by the United Nations. These agreements contain binding legal provisions covering a wide range of issues and provide an international legal basis for addressing specific environmental issues. In most cases, multilateral agreements are applied to solve regional environmental problems.

Active in international cooperation, China and the ASEAN Member States have signed and ratified a number of multilateral environmental agreements, entered into international conventions on environmental protection, and been fulfilling their obligations through implementation. As shown in Table 4.5, the existing multilateral agreements signed by China and the ASEAN Member States mainly cover biodiversity, water resources, land, chemicals and waste, and air.

The ratification of multilateral agreements reflects the varying degrees of concerns about environmental issues in countries. Because of rich biological resources, the ASEAN Member States have early ratified the Convention on Wetlands of International Importance and the Convention on Biological Diversity, reflecting the high attention paid to biodiversity. In recent years, China and more ASEAN Member States have ratified relevant international conventions on air and chemical wastes, expressing the determination to deal with transboundary wastes and atmospheric problems.

## 4.3 Typical Policy Measures Analysis

For the purpose of addressing global and regional environmental issues, the policies and programs that have been widely applied in the China and ASEAN are identified, followed by analysis of typical ones.

**Table 4.5** Major multilateral agreements participated by China and ASEAN Member States

| Biodiversity | Water | Land | Chemicals and waste | | | | | | | Air | | | |
|---|---|---|---|---|---|---|---|---|---|---|---|---|---|
| | Convention on Wetlands of International Importance | Convention on Biological Diversity (CBD) | Cartagena Protocol on Biosafety | Convention on International Trade in Endangered Species of Wild Flora and Fauna (CITES) | United Nations Convention on the Law of the Sea (UNCLOS) | United Nations Convention to Combat Desertification (UNCCD) | Stockholm Convention | Basel Convention | Rotterdam Treaty | Vienna Convention | Montreal Protocol | United Nations Framework Convention on Climate Change (UNFCCC) | Kyoto Protocol |
| China | 1992 | 1992 | 2005 | 1981 | 1996 | 1997 | 2004 | 1991 | 2005 | | | 1992 | 2002 |
| Brunei | Not signed | 2008 | | 1990 | 1996 | 2002 | | | Not signed | 1990 | 2009 | 2007 | 2009 |
| Cambodia | 1999 | 1995 | 2003 | 1997 | 1983 (Signed but not ratified) | 1997 | 2006 | | 2013 | 2001 | 2007 | 1995 | 2002 |
| Indonesia | 1992 | 1994 | 2004 | 1978 | 1982 | 1998 | 2009 | 1993 | 2013 | 1992 | 2006 | 1994 | 2004 |
| Laos | | 1996 | 2004 | 2004 | 1982 | 1996 | 2006 | | 2010 | 1998 | 2006 | 1995 | 2003 |
| Malaysia | 1995 | 1994 | 2003 | 1977 | 1982 | 1997 | | | 2002 | 1989 | 2001 | 1994 | 2002 |
| Myanmar | 2005 | 1994 | 2008 | 1997 | 1982 | 1997 | 2004 | | Not signed | 1993 | | 1994 | 2003 |
| The Philippines | 1994 | 1994 | 2004 | 1981 | 1982 | 2000 | 2004 | 1993 | 2006 | 1991 | 2006 | 1994 | 2003 |
| Singapore | | 1995 | | 1983 | 1982 | 2001 | 2005 | | 2002 | 1989 | | 1997 | 2006 |
| Thailand | 1998 | 2004 | 2005 | 1986 | | 1999 | 2005 | 1997 | 2005 | 1989 | | 1994 | 2002 |
| Vietnam | 1989 | 1994 | 2004 | 1994 | 1982 | 1998 | 2002 | | 2007 | 1994 | 1989 | 1994 | 2002 |

*Note* Blank is due to unavailability of information on whether a multilateral agreement is signed

### 4.3.1 Policy Measures to Address Climate Change

Climate change is a hot ecological and environmental issue in the world now. Global joint efforts and market-based instruments (MBI) are important pathways to address climate change. Therefore, this section analyzes the typical policy measures of multilateral environmental agreements and national plans, cancellation of the subsidy to fossil fuel, the feed-in-tariff of renewable energy and carbon emission trade.

#### 4.3.1.1 Multilateral Environmental Agreements

China and ASEAN have recognized the importance of addressing climate change issues and entered into international agreement under the principle of common but differentiated responsibilities (see Table 4.6). Signed in 1992, UNFCCC was the first agreement on climate change. China and ASEAN have no obligation emission reduction commitments, as non-Annex I parties to the Kyoto Protocol, a top-down agreement signed in subsequent years.

However, after reaching consensus at the UN Climate Change Conference in 2009, Indonesia, Malaysia, the Philippines, Singapore and China expressed their ambitions on climate change adaptation and mitigation by committing different extra emission reduction targets and goals officially (see Table 4.7).

The Paris Agreement, an outcome of the Paris Climate Change Conference 2015, was signed by 175 countries and regions on April 22, 2016 at the headquarters of the United Nations. China and ASEAN all signed the agreement and determined their own voluntary emission reduction targets. China as the biggest emitter has explicitly committed to, by 2030, cutting $CO_2$ emissions per unit of GDP by 60–65% from 2005 levels, raising the share of non-fossil energy to around

**Table 4.6** Participation in agreements on climate change

| Country | UNFCCC | Kyoto protocol | Paris agreement |
|---|---|---|---|
| Brunei | 2007 | 2009 | 2016 |
| Cambodia | 1995 | 2002 | 2016 |
| Indonesia | 1994 | 2004 | 2016 |
| Laos | 1995 | 2003 | 2016 |
| Malaysia | 1994 | 2002 | 2016 |
| Myanmar | 1994 | 2003 | 2016 |
| The Philippines | 1994 | 2003 | 2016 |
| Singapore | 1997 | 2006 | 2016 |
| Thailand | 1994 | 2002 | 2016 |
| Vietnam | 1994 | 2002 | 2016 |
| China | 1992 | 2002 | 2016 |

*Source* UNFCCC website and SoER4

**Table 4.7** Intended nationally determined contributions after the Copenhagen accord and the Paris agreement

| Country | Agreement | Target (%) | Remarks |
|---|---|---|---|
| Brunei | Post Paris agreement | 63 | Energy consumption, 2035 target, BAU |
| Cambodia | Post Paris agreement | 27 | 2030 target, BAU |
| Indonesia | Post Copenhagen accord | 26 | 2020 target, BAU |
| | Post Paris agreement | 29 | 2030 target, BAU |
| Malaysia | Post Copenhagen accord | 40 | 2005 baseline, 2020 target, energy intensity |
| | Post Paris agreement | 45 | 2005 baseline, 2020 target, energy intensity |
| the Philippines | Post Copenhagen accord | 20 | BAU for national emission growth |
| | Post Paris agreement | 70 | 2030 target, BAU |
| Singapore | Post Copenhagen accord | 16 | 2020 target, BAU |
| | Post Paris agreement | 36 | 2005 baseline, 2030 target, emission intensity |
| Thailand | Post Paris agreement | 20 | 2030 target, BAU |
| Vietnam | Post Paris agreement | 8 | 2030 target, BAU |
| China | Post Copenhagen accord | 40–45 | 2005 baseline, 2020 target, emission intensity |
| | Post Paris agreement | 60–65 | 2005 baseline, 2030 target, emission intensity |

*Source* ASEAN website, China's Policies and Actions for Addressing Climate Change (2011), UNFCCC

20%, peaking $CO_2$ emissions from fossil energy consumption around 2030, and increasing the forest stock by 4.5 billion $m^3$ from 2005 levels.

**Regional and national plans**

ASEAN has announced a regional plan to safeguard climate change policy implementation. The ASEAN Socio-Cultural Community Blueprint (2009–2015) specifies 11 strategic objectives and corresponding actions to mitigate climate change and its impacts. The Action Plan on Joint Response to Climate Change unveiled in 2012 develops a more specific blueprint for implementation.

At the national level, Indonesia, Laos, Malaysia, the Philippines, Singapore, Thailand, Vietnam and China have unveiled their national programs and strategies to address climate change. Brunei, Cambodia, and Myanmar announced specific projects which address climate change. Despite different national conditions, China and ASEAN have started to take a comprehensive policy on the issue of climate change (see Table 4.8).

**Table 4.8** Action plans and programs to address climate change

| Country | Action plans and programs |
|---|---|
| Brunei | • Study on national GHG inventories<br>• Energy efficiency improvement<br>• Renewable energy pilot projects<br>• Sustainable forest management<br>• Flood control and poverty reduction programs. |
| Cambodia | • Development of climate change policies, laws, policies, strategies and plans, including GHG mitigation and adaptation programs<br>• Promotion of policies related to climate change, integration of laws, policies, projects and programs, decision on national position and policy in international negotiations on climate change<br>• Coordination of relevant UNFCCC subsidiary agreements and other international climate change agreements<br>• Promotion of comprehension and dissemination of information on climate change |
| Indonesia | Develop policies, including the departmental measures on energy, transportation, agriculture, water and coastal management; National Climate Change Action Plan (2007) requires:<br>• Multi-sectoral (energy, transport, forestry and agriculture) coordination to further integrate emission reduction, climate change adaptation and national development goals<br>• Inclusion of climate-related funding decisions into all development plans, especially inclusion of the most promising system synergy into both medium-term and near-term development plan which is based<br>on the principle of "poverty alleviation to support economic growth and environmental protection" |
| Laos | National Adaptation Program of Action (NAPA); National Strategy and Action Plan Changes to Combat Climate Change (2009) requires:<br>• Improving measures to achieve the sustainable development goals of the Lao People's Democratic Republic, including low-carbon economic growth<br>• Building up capacity of key sectors of the national economy and natural resources to adapt to climate change and its effects<br>• Enhancing cooperation and building strong alliances and partnerships achieving national development objectives for national interests with international partners;<br>• Raising public awareness and understanding of different stakeholders on climate change and vulnerability impact, GHG emission source and relative contribution, as well as on climate change impact on the economy, in order to stimulate the enthusiasm of stakeholders about action |
| Malaysia | National Climate Change Policy (2009) requires:<br>• Mainstreaming climate change into intelligent resource management and environmental protection improvement, in order to improve economic strength and quality of life;<br>• Integration of national policies, programs and projects, in order to enhance the vitality of development under the potential impact of climate change<br>• Building up the institutional and implementation capacity in order to better take advantage of opportunities to reduce the negative impact of climate change |

(continued)

**Table 4.8** (continued)

| Country | Action plans and programs |
|---|---|
| Myanmar | • GHG inventories (Asian lowest-cost GHG abatement strategy)<br>• Preparation of national communications programs<br>• Implementation of the Clean Development Mechanism (CDM) in national institutions<br>• Preparation of national adaptation programs (NAPA)<br>• Sustainable forest management<br>• Improvement and promotion of energy efficiency |
| The Philippines | National Climate Change Action Plan 2011–2028 calls for priority to food security, adequate water supply, ecosystem and environmental stability, human security, climate-smart industries and services, renewable energy, and capacity building as strategic directions for strategic development directions during 2011—2028 |
| Singapore | National Climate Change Strategy 2012 requires:<br>• Public investment and information technology to help achieve a low-carbon transition<br>• Flexible supervision, effective policies and measures, and plans for the future |
| Thailand | National Strategy to Address Climate Change (2008–2012) and National Plan to Address Climate Change 2011–2050 require:<br>• Integrating all sectors into the response mechanism for adaptation to climate change and reduction of GHG emissions<br>• Stimulating enthusiasm of departments at all levels to developing and implementing plans to tackle climate change according to principles of relevance, appropriateness, effectiveness and specification<br>• Participating actively in international solution to climate change, by enhance sustainable development in the country<br>• Transiting to a sustainable, low-carbon society with economic and prosperity and self-sufficiency<br>• Making constant efforts to achieve effective and coordinated economic, social and environmental development<br>• Achieving by 2050, economic prosperity and GHG emissions without undermining GDP or development capacity and competitiveness |
| Vietnam | National Climate Change Strategy 2011 requires:<br>• Ensuring food security, energy security, water security, poverty reduction, gender equality, social security and public health, improving living standards, and protecting natural resources in the context of global climate change<br>• Mainstreaming low-carbon economy and green growth into the development<br>• Enhancing the awareness, responsibility and capability to tackle climate change, and taking full advantage of opportunities posed by climate change to promote social and economic development, and developing and extending climate-friendly consumption patterns<br>• Actively participating in international cooperation and enhancing effective actions to address climate change in order to contribute to climate change adaptation and mitigation |
| China | China's National Climate Change Program 2007 and China's Policies and Actions to Address Climate Change (2011) cover 11 areas and require:<br>• Accelerated transformation of economic development mode under the theme of scientific development |

(continued)

**Table 4.8** (continued)

| Country | Action plans and programs |
|---|---|
| | • Green low-carbon development, structural adjustment and transformation of economic development mode<br>• New industrialization, reasonable control of aggregate energy consumption. Measures taken include comprehensive optimization of industrial structure and energy mix, energy conservation and improvement of energy efficiency, and carbon sequestration capacity building |

*Source* SoER4, the EEA website, and the Chinese Central Government website

### Key policies for addressing climate change

The policies rolled out successively in China and ASEAN mitigate climate change, and the typical policies are categorized according to policy instruments. These successful experiences help to improve the capacity of the entire region in addressing climate change.

China adopts a target decomposition and assessment policy. National emissions reduction targets are decomposed and assigned to the provincial level, and the provincial governments are required to develop economic and social development plan and annual emissions reduction plans that ensure the realization of targets. The target of $CO_2$ emission intensity will be incorporated into economic and social development evaluation and government performance evaluation. Rewards and punishments will be given according to performance.

**Box 4.4 ASEAN actions on climate change**

The ASEAN Member States worked out a plan of actions to mitigate climate change, summarized as follows:

- On climate change adaptation and emission reduction, strengthen personal and institutional capacity building, in particular for disadvantaged and marginalized population;
- Promote comprehensive and consistent measures to address the challenge of climate change, including but not limited to the method of multi-stakeholders and multi-departments;
- Guarantee that private sectors and population can leverage on the new accessible and innovative financing mechanisms to address climate change;
- Strengthen the capacity of different departments and local governments in conducting GHG inventory and vulnerability assessment and meeting the needs;
- Reinforce the efforts of government, private sectors and communities in emission reduction from major development activities;
- Major climate change risk management and emission reduction planning by different departments;

- Strengthen global partnership, support the implementation of relevant international agreements and frameworks, e.g. UNFCCC.

Source: ASEAN Socio-Cultural Community Blueprint (2025), and the ASEAN website.

#### 4.3.1.2 Removal of Fossil Fuel Subsidies

China and six ASEAN Member States, i.e. Brunei, Indonesia, Malaysia, Myanmar, Thailand, and Vietnam, develop policy to lift fossil fuel subsidies, while introducing tiered pricing to a variety of energy sources (see Table 4.9) [1, 2].

#### 4.3.1.3 Feed-in Tariffs for Renewable Energy Projects

China, Indonesia, Malaysia, the Philippines, and Thailand implement a feed-in tariff scheme for renewable energy projects. According to the Interim Measures on Fund Management for Feed-in Tariffs for Renewable Energy Sourced Electricity, the Chinese Government invests 4000 yuan/kW-year to support renewable energy projects. Indonesia, Malaysia, the Philippines and Thailand also provide subsidies on renewable energy sourced electricity tariffs, of which Indonesia plans to raise subsidized prices in 2015 to encourage the development of renewable energy.

#### 4.3.1.4 Emissions Trading

CDM is a flexible mechanism under the Kyoto Protocol (IPCC, 2007) that allows the trade of certified emission reductions (CER) units generated by emission reduction projects. China and ASEAN, except for Brunei, have participated to varying degrees in CDM projects, of which noticeably active are China, Malaysia, Thailand and Indonesia. As non-Annex I countries, China and ASEAN can contribute to tackling climate change by developing CDM projects (Table 4.10).

China is actively pressing ahead with the emissions trading pilot. In 2014, the pilot was kicked off in Beijing, Shanghai, Tianjin, Guangdong, Shenzhen, Chongqing, and Hubei, involving about 12 million tons allowances for more than 1900 enterprises. China intends to launch a national emissions trading market in 2017, which will make China the first developing country to embrace a national carbon market. It is predicted that by 2020 China will become the world's largest emissions trading market.

**Table 4.9** Removal of fossil fuel subsidies in China and ASEAN

| Country | Subsidized products | Measures |
|---|---|---|
| Brunei | Gasoline, diesel, liquefied petroleum gas (LPG), and electricity | Increase gasoline and diesel prices in 2008, in order to restrict "fuel tourism" of vehicles from Malaysia, and raise the prices for foreign cars again in 2012 |
| Indonesia | 88-octane gasoline (for small businesses and residents), diesel, kerosene; LPG, and electricity | Raise the gasoline price by 44% and diesel price by 22% in June 2013. Promote the use of natural gas in the transportation sector in order to reduce oil subsidies. Continue the successful kerosene-to-LPG plan that began in 2007. Raise the electricity tariff will rise by 15% in 2013 (on a quarterly basis) |
| Malaysia | 95-octane gasoline, diesel, LPG, and electricity | Gradually raised gas and electricity prices in 2011 and planned to cancel subsidies in 2014. Subsidies for gasoline and diesel were cut in September 2013 in order to achieve deficit reduction |
| Myanmar | Electricity, gas, and kerosene | Link diesel and gasoline prices to the Singapore spot market in 2011. Raise electricity price in January 2012 as a part of the reform of the electricity system |
| Thailand | LPG price control, price control of diesel and natural gas (for vehicles) in order to reduce the impact of international price volatility, and price control of electricity for poor families | Raise LPG prices on a monthly basis from September 2013 onwards, while ensuring that the lowest burden of electricity consumption on street vendors and consumers. Raise the prices every four month since September 2013 |
| Vietnam | Diesel, gasoline, natural gas, and electricity | Gradually move toward market-oriented oil and gas prices. Plan to launch a road map to remove fossil fuel subsidies |
| China | Petroleum products, natural gas | Implement tiered pricing system in July 2012. Announce in March 2013 to adjust the prices of petroleum products every 10 days to better reflect changes in the international oil market. Raise the price of non-residential natural gas by 15% in July 2013, accounting for about 80% of natural gas demand |

*Source* the International Energy Agency (IEA) website and the IEA ASEAN Report

**Table 4.10** Registered CDM projects (as of May 2014)

| Country | Number of projects | Total emission reductions (Metric ton) |
|---|---|---|
| Brunei | None | None |
| Cambodia | 10 | 2,021,187 |
| Indonesia | 146 | 17,823,691 |
| Laos | 9 | 1,303,092 |
| Malaysia | 143 | 8,404,905 |
| Myanmar | 1 | 709 360 |
| The Philippines | 72 | 3,559,240 |
| Singapore | 4 | 179,979 |
| Thailand | 145 | 7,372,252 |
| Vietnam | 252 | 17,876,191 |
| China | 3759 | 596,432,055 |

*Source* UNFCCC CDM website

## 4.3.2 Policy Measures for Biodiversity Conservation

Biodiversity conservation is a key part of ecological and environmental issue. It may involve many policies, but still is dominated by command-and-control policy and market-based instruments as the major policy categories. This section discusses the major policy measures of multilateral environmental agreements and national plan, ecological compensation and habitat management.

### 4.3.2.1 Multilateral Environmental Agreements

Except that Singapore and Brunei did not participate in the Convention on Wetlands of International Importance and the Cartagena Protocol on Biosafety, China and ASEAN haven entered at different times into all multilateral agreements on biodiversity and ecosystem conservation (see Table 4.11), including the 1971 Convention on Wetlands of International Importance (Ramsar Convention), 1972 Convention Concerning the Protection of the World Cultural and Natural Heritage, CITES 1975, CBD 1992, 2000 Cartagena Protocol on Biosafety, and UNCCD 1994.

**Regional and national plans**

ASEAN has declared a regional plan to ensure implementation of policies on biodiversity and ecosystem management policies. For example, the ASEAN Socio-Cultural Community Blueprint (2009–2015) listed a number of strategic actions to enhance sustainable management of natural resources and biodiversity.

**Table 4.11** Entry into multilateral environmental agreements related to biodiversity

| Country | CBD | Ramsar convention | World Heritage convention | CITES | Cartagena protocol on biosafety | UNCCD |
|---|---|---|---|---|---|---|
| Brunei | 2008 | None | 2011 | 1990 | None | 2002 |
| Cambodia | 1995 | 1999 | 1991 | 1997 | 2003 | 1997 |
| Indonesia | 1994 | 1992 | 1989 | 1978 | 2004 | 1998 |
| Laos | 1996 | 2010 | 1987 | 2004 | 2004 | 1996 |
| Malaysia | 1994 | 1995 | 1988 | 1977 | 2003 | 1997 |
| Myanmar | 1994 | 2005 | 1994 | 1997 | 2008 | 1997 |
| The Philippines | 1993 | 1994 | 1985 | 1981 | 2006 | 2000 |
| Singapore | 1995 | None | 2012 | 1986 | None | 1999 |
| Thailand | 2004 | 1998 | 1987 | 1983 | 2005 | 2001 |
| Vietnam | 1994 | 1989 | 1987 | 1994 | 2004 | 1998 |
| China | 1993 | 1992 | 1985 | 1981 | 2005 | 1997 |

**Box 4.5 ASEAN actions on biodiversity and ecosystem management**

- Achieve by 2010, a significant reduction in the current rate of loss of biodiversity through implementing relevant national, regional and international programs of work;
- sharing of genetic and biological resources by 2015;
- Promote further listing and coordinated management of ASEAN Heritage Parks as an effective platform for ecosystem-based protected areas management by 2015;
- Enhance cooperation in the management of transboundary protected areas between neighboring ASEAN Member States (AMS);
- Take appropriate measures to minimize impacts of transboundary movement of living modified organisms in accordance with the Cartagena Protocol on Biosafety by 2015;
- Establish a functional regional network to promote capacity building in developing inventory of the biological resources and biosafety measures of the ASEAN Region by 2015;
- Enhance the role and capacity of the ASEAN Centre for Biodiversity (ACB) to function as an effective regional centre of excellence in promoting biodiversity conservation and management;
- Promote the involvement of local community to maintain biodiversity conservation and forest health by 2015;
- Promote effective management policies and practices to reduce the impact of invasive alien species at the regional and international levels;
- Promote regional cooperation on sustainable management of biodiversity such as sharing research and development experiences, exchange of experts, and training;

- Strengthen efforts to control transboundary trade in wild fauna and flora through the ASEAN Action Plan on Trade in Wild Fauna and Flora 2005–2010 and the ASEAN Wildlife Enforcement Network (ASEAN-WEN) to implement CITES commitments;
- Explore cooperation among AMS to conduct joint survey and monitoring of migratory wildlife;
- Promote cooperation among AMS in combating land degradation for stainable land management in support of sustainable agriculture and environmental protection.

Source: ASEAN Socio-Cultural Community Blueprint (2009–2015), and the ASEAN website.

At the sub-regional level, the Greater Mekong sub-regional cooperation has been carried out for many years. Countries in the GMS cooperation, i.e. China, Myanmar, Laos, Thailand, Cambodia and Vietnam should protect their favorable environment and rich biodiversity in the cooperative development process. Protection forests are needed to control soil erosion in major river basins, including Lancang-Mekong River, Red River, Irrawaddy River and Salween River, and meanwhile, the nature conservation network should be established for protection of rare species and effective control of environmental pollution.

At the national level, China and ASEAN Member States all issued National Biodiversity Strategies and Action Plans (NBSAPs) (see Table 4.12). Indonesia, the Philippines, Thailand, and China have also updated their NBSAPs which basically covered the priority actions and a few define the key principles of ecological protection.

**Biodiversity and ecosystem conservation policy**

Rich in biological diversity, China and the ASEAN Member States have implemented many policies to enhance biodiversity and ecosystem management. A number of successful policy measures can be identified and extended to other member states to drive improvement in the whole region [3].

#### 4.3.2.2 Ecological Compensation

China and Vietnam adopt the policy of ecological compensation. China identifies ecological compensation as a major national environmental policy aimed at promoting sustainable use of natural resources and more balanced growth among regions. China has implemented many of the world's largest ecosystem service payment programs. For example, the expenditure on reforestation from 1999 onwards has exceeded 15 billion US dollars, which has brought trees and grass to over 9 million ha of slop arable land. In this program, farmers are compensated for

**Table 4.12** National Biodiversity Strategies and Action Plans (NBSAPs)

| Country | National strategies and action plans | | | | |
|---|---|---|---|---|---|
| | NBSAP | Area | Priority action | Priority program | Remarks |
| Brunei | None | | | | |
| Cambodia | 2002 | 17 | 98 | | Its NBSAP suggests establishment of biodiversity management department |
| Indonesia | 1993 (2003) | 5 | 72 | | Its NBSAP is targeted at time frame during 2003–2020 and puts emphasis on partnership, decentralized and transparent biodiversity management |
| Laos | 2004 | 7 | 27 | 201 | Its NBSAP includes the strategy toward 2020 and action plans toward 2010. It sets seven main objectives, including identification of important biodiversity components and improvement of knowledge base |
| Malaysia | 1998 | 11 | 15 | | The guiding principle of the national policy is "ensuring the role of local communities and reasonable benefit-sharing in the protection, management and use of biological diversity" |
| Myanmar | 2006 | 5 | 20 | 162 | A three-tier structural system is identified at the initial stage and the specific institutional structure is to be formed |
| The Philippines | 1997 | 5 | 206 | | The Philippines Biodiversity Conservation Priority Program has identified 206 priority areas for protection |
| Singapore | 1992 (2002) (2009) | 5 | 17 | | Its NBSAP integrates more detailed biodiversity strategies and efforts of the government, communities and organizations in coming years than the Singapore Green Plan |
| Thailand | 1997 (2002) (2008) | 7 | 26 | | Its second NBSAP covers efforts in seven thematic areas during 2003–2007 |
| Vietnam | 1994 (2007) (2015) | 5 | 16 | | A new national biodiversity action plan was approved in 2007. It highlighted the legislation for biodiversity. Biological Diversity Act was promulgated in 2008. The legal framework and management system have been completed |
| China | 1994 | 7 | 26 | 18 | China's NBSAP was issued in 2010, covering |
| | (2010) | (10) | (39) | (39) | 10 key areas, 30 prioritized action plans and |
| | | | | | 39 prioritized action programs |

() means the updated years
*Source* SoER4, and the Chinese MEP website

change in use of agricultural land. Similarly, China has allocated nearly 2 billion US dollars for the forest ecosystem compensation fund, and paid families, communities and local governments to protect critical forest areas which to date reaches nearly 44 million ha. Vietnam has successfully implemented the demonstration project on payments for forest environmental services in Lam Dong Province, with assistance of the Regional Development Mission for Asia of the United States

Agency for International Development (USAID). Apart from biodiversity conversation, the project has improved the life of 40,000 rural poor people and reported the design and follow-up of national act associated the program. The Act created the legal framework for integrated ecosystem management in experimental Lam Dong and Son La provinces. The move not only promotes payments for forest management, but also increases the revenue of service-providing communities [4].

**Box 4.6 Ecological compensation in Lam Dong Province of Vietnam and Xin'anjiang of China**

The project of payments for ecosystem services was implemented in South Vietnam's Lam Dong Province, in which forest service buyers, referring to electricity and water utilities, pay for water diversion and soil conservation and tourism operators bear the landscaping cost. As of December 2012, the charges for protection of 210,000 ha of forests, involving more than 4 million US dollars, have been given to 22 forest administrations and forestry companies and 9870 major ethnic minority families with an annual average of 540–615 US dollars per household. Forest protection patrols supported by the project have halved the occurrence of illegal logging and poaching in protected areas in the province. The dissemination of such compensation practice in Vietnam will exert a major effect by promoting forest habitat preservation and biodiversity conservation, especially when the number of buyers of ecosystem services increases.

In 2012, China kicked off the pilot project for ecological compensation mechanism in Xin'anjiang to have the ecological benefitting communities to compensate those who suffered losses. The pilot compensation plan requires an annual compensation fund of 500 million yuan, with 300 million from central budget and 100 million respectively by Anhui and Zhejiang provinces. The major monitoring indicators for water quality is potassium permanganate, ammonia nitrogen, total nitrogen, total phosphorus and the monitoring evidence is largely from the water body section across the two provinces. Ever since the implementation of the pilot, the local ecological environment has been improved gradually and the industrial structure is also optimized [5].

#### 4.3.2.3 Habitat Management

China and ASEAN implement a protected area policy for habitat administration. ASEAN classifies protected areas into three categories: (a) nature reserves, wilderness areas and national parks; (b) protected natural habitats and terrestrial and marine protected areas for species management; and (c) protected areas for sustainable use of natural resources. For critical areas of biodiversity, ASEAN Heritage

Parks and transboundary protected areas, ASEAN develops biodiversity protection plans and ecosystem management. In China, the protected areas are divided into (a) typical natural geographical areas; (b) natural concentrated distribution areas; (c) areas of rare and endangered wildlife species; (d) areas of particular conservation value (scientific and cultural values); and (e) government-approved protected areas. At present, there are more than 2500 protected areas in China.

**Box 4.7 ASEAN Turtle Islands Heritage Protected Area**

Turtle Islands Heritage Protected Area (TIHPA) is the first transboundary protected area in the world, and its area of coverage spans Malaysia and the Philippines. It is the major nesting ground of the green sea turtle (Chelonia mydas) and is the only remaining nesting rookery of green sea turtles in the ASEAN region. It is also the eleventh major nesting area of marine turtles in the world. Turtles lay hundreds of thousands of eggs in the TIHPA each year, with approximately more than 2000 nesters. Hawksbill turtle (Eretmochelys imbricata) also nests in the TIHPA.

Biodiversity resources in the TIHPA include 34 avian species, fruit and field bats, and reptiles (snakes, green sea and hawksbill turtles, monitor lizards); 15 principal arborescent species; 24–27 genera of corals; 76–128 fish species; and 62 species of marine flora.

### *4.3.3 Policy Measures to Protect Freshwater Resources*

Freshwater resources are key component of regional ecological and environmental concerns. On both water quantity and quality, command-and-control policy and market-based instruments (MBI) are still major policy categories. This section discusses the typical policy measures water resource regional plan, river basin management planning, tiered water pricing policy, pollution discharge permit, and mandatory environmental protection fee.

#### 4.3.3.1 Water Resource Regional Plan

ASEAN has announced a regional plan to ensure the implementation of policy on freshwater resources. A total of 13 actions to enhance the sustainability of freshwater resources are provided in the ASEAN Socio-Cultural Community Blueprint 2009–2015.

**Box 4.8 ASEAN actions on freshwater resources**

- Continue implementation of the ASEAN Strategic Plan of Action on Water Resources Management;
- Endeavor to reduce by half the number of people without sustainable access to safe drinking water by 2010;
- Manage water resources efficiently and effectively in order to provide adequate and affordable water services by 2015;
- Promote the implementation of integrated river basin management by 2015;
- Promote public awareness and partnership to enhance integrated water resources management;
- Promote regional cooperation on water conservation measures and program as well as scientific and technological innovations in water quality improvement and supply.

Source: ASEAN Socio-Cultural Community Blueprint 2009–2015, and the ASEAN website.

In the field of water resources management, ASEAN issued in 2005 the ASEAN Strategic Plan of Action on Water Resources Management which marks a landmark achievement of environmental cooperation in the region. The plan introduces the world's best practices in integrated water resource management and capacity building, with the purpose to improve the community's understanding of importance of water resources [6].

#### 4.3.3.2 River Basin Management Planning

China adheres to integrated river basin management and administration under the water resource management system. Malaysia, Cambodia, the Philippines, and Indonesia implement a decentralized management system, which defines the responsibilities of governments and districts at all levels. In Thailand, the river basin agencies play a considerable role in water resource management. A license and authorization system for groundwater development is set up in Thailand, Myanmar, Laos and Vietnam.

#### 4.3.3.3 Tiered Pricing Policy

China, Malaysia, the Philippines and Vietnam introduce water resource fees. Malaysia imposes affordable fees on low-income groups and full water charges on high-income group. In the Philippines and Vietnam, water charges have been adjusted to reflect the economic value of water resources. China is implementing a progressive water tariff to control water consumption.

#### 4.3.3.4 Pollution Permit System

China implements emission standards and total emission control for the discharge of pollutants. Individuals and organizations that discharge pollutants to the environment shall apply in advance for and obtain the corresponding pollution permit from environmental protection departments. The emissions standards for water pollutants put differentiated caps of pollutant concentration for industries. The emission allowances for aggregate water pollutants are developed and decomposed by the Central Government and made known to enterprises through local governments.

China, the Philippines and Singapore impose taxes/charges on water pollutant emissions. In the Philippines, companies are required to pay a fixed fee based on the amount of wastewater emissions and the type and quantity of water pollutants. In Singapore, conventional water and electricity charges and water consumption tax are joined by water fees. In China, pollution charges will be doubled where emissions are beyond the standards, taking into account of the type and quantity of pollutants.

#### 4.3.3.5 Compulsory Environmental Taxes and Fees

Environmental fees are made mandatory in Vietnam. Ho Chi Minh City imposes a 4% resource tax on water consumption, which equals to 160 VND/$m^3$ for groundwater and 80 VND/$m^3$ for surface water, according to prices of 4000 and 2000 VND/$m^3$ respectively.[4]

### *4.3.4 Policy Measures for Air Pollution Prevention and Control*

Air pollution is an important issue that affects human health and ecosystem and features transboundary movement and thus calls for regional management. This section discusses the typical policy measures of multilateral environmental agreements, air pollution action plan, and air pollution emission cap control.

---

[4]Source: Economic and Commercial Section of the Consulate General of the People's Republic of China in Ho Chi Minh City, 2007, http://hochiminh.mofcom.gov.cn/aarticle/jmxw/200707/20070704844763.html.

**Table 4.13** Participation in multilateral environmental agreements on air pollution

| Country | Montreal protocol | Vienna convention |
|---|---|---|
| Brunei | 1993 | 1990 |
| Cambodia | 2001 | 2001 |
| Indonesia | 1992 | 1992 |
| Laos | 1998 | 1998 |
| Malaysia | 1989 | 1989 |
| Myanmar | 1993 | 1993 |
| The Philippines | 1991 | 1991 |
| Singapore | 1989 | 1989 |
| Thailand | 1989 | 1989 |
| Vietnam | 1994 | 1994 |
| China | 1991 | 1989 |

#### 4.3.4.1 Multilateral Environmental Agreements

Recognizing the importance of addressing air pollution, China and ASEAN have entered into the multilateral agreements, including the 1985 Montreal Protocol on Substances that Deplete the Ozone Layer and the 1987 Vienna Convention for the Protection of the Ozone Layer (see Table 4.13).

#### 4.3.4.2 Action Plans

The measures to control air pollution, a common issue facing China and ASEAN, vary among countries (see Table 4.14) [7]. The focus of air pollution control depends on the degree of pollution and level of economic and social development.

The ASEAN Agreement on Transboundary Haze Pollution was signed in 2002 in Kuala Lumpur and entered into force in 2003. The agreement aims to improve management policies and enforcement mechanisms for preventing land and/or forest fires, establish operational mechanisms for monitoring land and/or forest fires, and to strengthen regional capacity and measures to fight land and/or forest fires. All ASEAN Member States ratified the Agreement (see Table 4.15).

#### 4.3.4.3 Control of Aggregate Air Pollutants

China implemented a total emission control system. Based on pollution permit system, enterprises and institutions, especially those tasked with control of total pollutant emissions, shall discharge pollutants as required in permitted conditions. New and expanded thermal power plants and other large- and medium-sized enterprises, are required to install facilities or take other measures for desulphurization and dust removal where the emissions fall short of emissions standards or control requirements.

**Table 4.14** Typical regulations and programs on air pollution control

| Country | Policy category | Regulations and programs |
|---|---|---|
| Indonesia | Others | 1996 Blue Sky Program |
| Malaysia | Oil products | Environmental Quality (Control of Petrol and Diesel Properties) Regulations 2007 and Environmental Quality (Clean Air) Regulations 1978, Environmental Quality (Control of Emission from Petrol Engines) Regulations 1996 and Environmental Quality (Control of Emission from Diesel Engines) Regulations 1996 pending approval. Clean Air Action Plan is also being developed to encourage all sectors to improve air quality |
| The Philippines | Oil products | Biofuels Act of 2006 approved on November 29, 2006, stipulates a minimum of one percent (1%) blend of biodiesel by volume into all diesel fuel distributed and sold in the country, and a minimum of five percent (5%) bend of bioethanol to all gasoline fuel<br>Provision X in DAO 2000-81 Section 5 and relevant provisions (DAO 2007-22) were adjusted and clarified in the continuous emission monitoring system and agreement guidelines |
| Thailand | Comprehensive | Air Quality and Noise Management in Thailand Master Plan B.E. (2005–2016) is drafted by Pollution Control Department (PCD). The $NO_2$ and noise standards under the Ambient Air Quality Standards (2007–2008) were published in order to improve air quality and reduce noise pollution. |
| Vietnam | Vehicle | The roadmap to application of on-road motor vehicle emission standards was approved in 2005. According to this roadmap, the Euro II standards applied to imported second-hand motor vehicles in July 2006 and all domestic and imported cars in July 2007. By 2025 year-end, Euro V standards will be applied to all vehicles in Viet Nam. |
| China | Comprehensive | Law of the People's Republic of China on Air Pollution Prevention and Control (2000)<br>Action Plan for Air Pollution Prevention and Control (2013) |

*Source* SoER4

**Table 4.15** Ratification of ASEAN Agreement on Transboundary Haze Pollution

| Country | Year of ratification |
|---|---|
| Brunei | 2003 |
| Cambodia | 2006 |
| Indonesia | 2014 |
| Laos | 2004 |
| Malaysia | 2002 |
| Myanmar | 2003 |
| The Philippines | 2010 |
| Singapore | 2003 |
| Thailand | 2003 |
| Vietnam | 2003 |

### 4.3.5 Policy Measures for Chemicals and Waste Management

The issue of chemicals and waste management emerged with population growth and economic development. The present policy measures are still dominated by command- and-control approach. This section discusses the typical policy measures of national laws, multilateral environment agreement, urban waste treatment management and hazardous chemical registration management.

#### 4.3.5.1 National Laws

China and ASEAN have developed a number of regulations and plans on chemicals and waste management, as shown in Table 4.16.

#### 4.3.5.2 Multilateral Environmental Agreements

China and most ASEAN Member States have entered into the multilateral environmental agreements on chemicals and waste management (except that Brunei does not join the Rotterdam Convention and Myanmar does not join the Basel Convention and Rotterdam Convention) (see Table 4.17). The major multilateral environmental agreements in this field include the Basel Convention on the Control of Transboundary Movements of Hazardous Wastes and Their Disposal (1989), Rotterdam Convention on the Prior Informed Consent Procedure for Certain Hazardous Chemicals and Pesticides in International Trade (1998), and Stockholm Convention on Persistent Organic Pollutants (2001).

#### 4.3.5.3 Urban Solid Waste Treatment Management

China implements the urban solid waste charge policy. In line with the polluter pays principle, waste producers are required to pay for urban solid waste treatment, considering treatment costs and national revenue, and the charges shall be used for urban solid waste collection and treatment.

Malaysia and the Philippines reconstruct urban waste landfills. There are both urban waste landfills and open dumps in the two countries. To reduce the impact on environment, a part of open dumps are identified to be rebuilt into municipal waste landfills.

**Table 4.16** Typical national regulations and plans on chemicals and waste management

| Country | Regulations and plans |
|---|---|
| Cambodia | Law on Environmental Protection and Natural Resources Management |
| Indonesia | Act of the Republic of Indonesia Number 18 Year 2008 about Waste Management |
| Laos | Environmental Protection Law of Lao People's Democratic Republic 1998: (a) Regulation on the Monitoring and Control of Waste Discharge (No. 1122/STENO) 1998; (b) Guidelines for Hospital Waste Management (1997); (c) Regulation on Industrial Wastewater Discharge (No.180/MIH) 1994; (d) policy and program compliance evaluation and monitoring |
| Malaysia | Environmental Quality (Scheduled Wastes) Regulations 2005<br>Solid Waste and Public Cleansing Management Act 2007<br>Solid Waste and Public Cleansing Management Corporation Act 2007 |
| The Philippines | Toxic Substances and Hazardous and Nuclear Waste Control Act (R.A. 6969) |
| Singapore | Environmental Protection and Management (Air Impurities) Regulations 2001 |
| Thailand | Enhancement and Conservation of National Environmental Quality Act 1992<br>Public Health Act 1992<br>Public Cleansing Act 1992<br>Building Control Act 1992<br>Factory Act 1992<br>Hazardous Substances Act 1992<br>Industrial Estate Authority of Thailand Act 1979 |
| Vietnam | (a) Land Law (completed in 1989, 1993); (b) Forest Protection and Development Law (1991); (c) Human Health Protection Act (1989); (d) Mineral Resources Act (1989); (e) Aquatic Products Act (1988) |
| China | Law of the People's Republic of China on Solid Waste Pollution Prevention and Control, 2004<br>Hazardous Chemicals Regulations, 2011<br>Measures for Municipal Solid Waste Management, 2007 |

**Table 4.17** Participation in multilateral environmental agreements on chemicals and waste management

| Country | Basel convention | Rotterdam Convention | Stockholm convention |
|---|---|---|---|
| Brunei | 2012 | None | 2002 |
| Cambodia | 2001 | 2013 | 2006 |
| Indonesia | 1993 | 2013 | 2009 |
| Laos | 2010 | 2010 | 2006 |
| Malaysia | 1993 | 2002 | 2002 |
| Myanmar | None | None | 2004 |
| The Philippines | 1993 | 2006 | 2004 |
| Singapore | 1996 | 2005 | 2005 |
| Thailand | 1997 | 2002 | 2005 |
| Vietnam | 1995 | 2007 | 2002 |
| China | 1991 | 2005 | 2004 |

#### 4.3.5.4 Hazardous Chemicals Registration Management

China's registration measures apply to hazardous chemicals, covering mainly three stages:

Production and storage—According to the law on industrial solid waste registration, producers shall update the industrial solid waste information to local environmental protection departments. According to the hazardous chemicals registration law, producers shall apply for safety production license, and hazardous chemicals shall be labeled. Producers are also required to update the information on inspection and registration of hazardous chemicals.

Use and management—The users of hazardous chemicals are required to apply for safety use license, and hazardous chemicals companies to apply for safety business license. To ensure safety, enterprises are prohibited from hazardous chemicals business without license.

Transportation—The inter-provincial transportation of solid waste shall receive the prior approval of environmental protection department of the source province, otherwise it shall be prohibited. Solid waste that cannot be reused are not allowed to import. In China, dedicated permits are required for the transportation of hazardous chemicals and road transportation of toxic chemicals, and the waterway transportation of toxic chemicals is prohibited.

### *4.3.6 Policy Measures to Protect Coastal Resources*

China and ASEAN Member States enjoy massive costal resources. This section discusses the policies of multilateral environmental agreements, regional plan and comprehensive costal management.

#### 4.3.6.1 Multilateral Environmental Agreements

China and ASEAN except for Cambodia have joined the UNCLOS (see Table 4.18).

#### 4.3.6.2 Regional Plans

ASEAN has declared a regional plan to ensure the implementation of policies on protection of coastal and marine resources. In specific, the ASEAN Socio-Cultural Community Blueprint 2009–2015 lists a number of actions to promote sustainable use of coastal and marine environment.

**Table 4.18** Entry into multilateral environmental agreements on coastal resource conservation

| Country | UNCLOS |
|---|---|
| Brunei | 1996 |
| Cambodia | None |
| Indonesia | 1986 |
| Laos | 1998 |
| Malaysia | 1996 |
| Myanmar | 1996 |
| The Philippines | 1984 |
| Singapore | 1994 |
| Thailand | 2011 |
| Vietnam | 1994 |
| China | 1996 |

**Box 4.9 ASEAN actions on coastal and marine environmental protection**

- Enhance inter-agency and inter-sectoral coordination at the regional and international levels for achieving sustainable development of ASEAN's coastal and marine environment;
- Build capacities to develop national marine water quality standards by 2015 using the ASEAN Marine Water Quality Criteria as a reference;
- Establish a representative network of protected areas to conserve critical habitats by 2015 through further implementation of the ASEAN Criteria for Marine Heritage Areas and ASEAN Criteria for National Protected Areas;
- Promote conservation and sustainable management of key ecosystems in coastal and marine habitats, such as joint efforts to maintain and protect marine parks in border areas and the "Coral Triangle Initiative on Coral Reefs, Fisheries and Food Security";
- Enhance the capacity and capability of, as well as economic benefits for the fishery and other coastal community to encourage their active participation in promoting environmental sustainability;
- Promote the sustainable use of coastal and marine environment through public awareness campaign to highlight the global importance of coastal and marine environment in addressing food security, maintaining ecosystem services, as well as protecting marine environment;
- Promote collaboration among ASEAN Member States in responding to transboundary pollution due to the oil spill incidents;
- Promote cooperation in addressing pollution of coastal and marine environment from land-based sources.

Source: ASEAN Socio-Cultural Community Blueprint 2009–2015, the ASEAN website.

#### 4.3.6.3 Integrated Coastal Zone Management

China and ASEAN create marine protected areas (MPAs). In China, MPAs are classified into four categories: (a) typical marine natural geographical areas, (b) representative natural ecological areas with highly rich marine species or naturally concentrated rare and endangered marine species, (c) sea area, sea coast, islands, coastal wetlands, estuaries and bays of particular conservation value, and (d) locations of marine natural heritage of great scientific and cultural value. For better MPA management, ASEAN Member States have carried out policy improvement and legislative reform, including MPA planning and management to achieve integrated coastal zone management framework and enhanced mechanisms that enable managers and institutions to continue adaptive management for adequate representation while overcoming the differences in understanding of biogeographic area establishment, and to improve joint research and cooperative management.

### *4.3.7 Policy Measures to Address Soil Degradation and Soil Erosion*

The governance policies of China and ASEAN on soil are largely command-and-control policies. This section discusses the policies on comprehensive management of agriculture, and water and soil erosion.

#### 4.3.7.1 Comprehensive Management of Agriculture

China takes comprehensive measures to control agricultural non-point source pollution. The measures include regulation of chemical fertilizers, pesticides and plant growth hormone, recycling of agricultural film, disposal of poultry manure, and treatment of wastewater and other waste.

Malaysia and Indonesia adopt a sustainable palm oil policy. The policy emphasizes the importance of environmental protection and rests on credible international standards and principles while taking into account the well-being of local communities and plantation workers.

#### 4.3.7.2 Soil and Water Erosion Policy

China's National Soil and Water Conservation Plan (2015–2030) sets 23 state-level key areas of water and soil erosion prevention which are categorized into important river source areas, important water source areas, and water-wind erosion crisscross areas and 17 key areas of water and soil loss control including northeast rolling hill areas. Focusing on areas in most urgent need of protection and treatment, a number of key prevention and treatment projects are designed. In state-level key areas of water and soil erosion prevention, construction projects and production activities are strictly controlled, and overgrazing, disorderly mining, deforestation and grassland reclamation are prohibited, in order to prevent soil and water loss and ecological damage from the source. Where there is water and soil erosion, comprehensive treatment is carried out according to local conditions. Efforts also include clean small watershed construction in important water source areas and special treatment of sloping cultivated land and ditch intensive areas.

## 4.4 Policy Evaluation

The benefits of certain policy/policies are reviewed in a number of evaluation reports despite the absence of elaboration on policy effects in China and ASEAN. Therefore, this section will not systematically expound the benefits of policy options, but only provide a general analysis.

### 4.4.1 Economic Benefits

The policy measures taken by China and ASEAN have generated two major results: businesses are mobilized to treat pollution; while the industrial structure is optimized, so that the resources are more effectively allocated to eco-friendly business. Preferential taxes and subsidies for energy efficient and pollution control products and acts also guide businesses and consumers to pollution prevention and control. By driving the development of advanced control technologies and environmental protection industries, these measures favor rational resource allocation and industrial structure, encourage business transformation, and expand employment opportunities. Enterprises that cause environmental pollution and damage are faced with pollution charges and fees and environmental taxes which will be used for pollution control and environmental protection. This move not only internalize the external costs of enterprises and rationalize profit levels, but also reduce the tax burden on businesses in compliance with environmental requirements, which better embodies the principle of fairness conducive to fair competition among enterprises of all kinds.

### 4.4.2 Social Benefits

The social impact of policy measures covers access to information, stakeholders, law enforcement, and gender equality, as well as environmental justice, equality and benefit-sharing, all of which will facilitate the transition to sustainable development. The compensation for the poor, such as ecological compensation to encourage poor areas to reforest, can reduce the number of population in poverty while protecting the environment, and therefore can be seen as a component of effective policy portfolio. The main obstacles to policy measures include unequal rights, conflicting social and personal values, and prevailing non-green consumption. The internalization of environmental costs incurred in the production and consumption process easily leads to rise of family costs and undermines employment in polluting sunset industries.

### 4.4.3 Environmental Benefits

In order to reduce poverty, China and ASEAN are committed to economic development in recent years by accelerating the pace of urbanization. A lot of environmental problems have arisen in this process, but the ecological environment in the region does not suffer noticeable impact owning to a series of policy measures. Policy interventions have produced significant environmental benefits. Environmental Performance Index (EPI) quantifies and numerically marks the

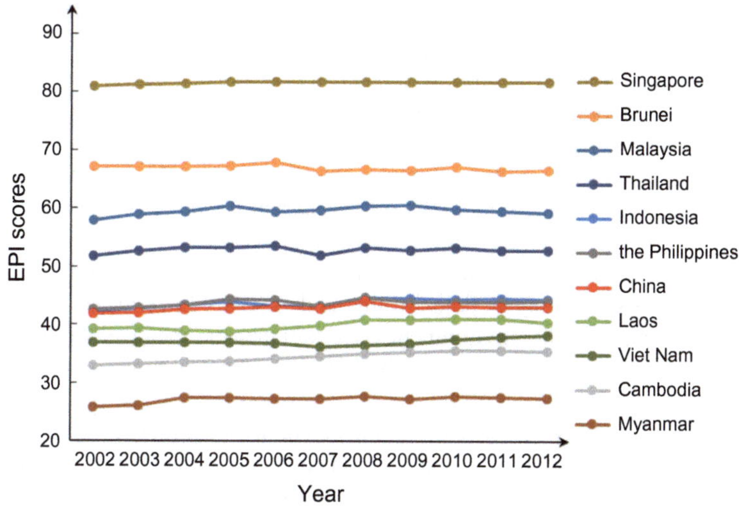


**Fig. 4.2** EPI scores of China and the ASEAN countries *Source* Yale University, http://www.epi.yale.edu/

environmental performance of a state's policies and reflects the overall progress of a country or region in the field of resource and environment. According to the EPI serial data released by Yale University every two years since 2004, the EPI rankings of China and the ASEAN Member States spanned from 4 to 164 in 2012, implying large disparity in environmental performance. Yet, the EPI scores were relatively stable during 2002–2012 (see Fig. 4.2), which would have been impossible if without such policy interventions, and regional environmental problems could have been more serious.

### *4.4.4 Policy Limitations and New Policy Demands*

Although the environment intervention policies of China and ASEAN have achieved very good economic, social and environmental benefits, due to economic development level, traditional management concept and the understanding toward the issue of environment, the policy interventions are still restricted by the following aspects:

First, policy design. Environmental problems accompany economic development and attract more attention as they become serious. In the face of a number of environmental problems arising from economic and social development, China and ASEAN can learn from the experience of other countries. However, a specific policy, if not appropriately designed in the dissemination process to accommodate specific context, may fail due to different national conditions. It is therefore necessary to carefully carry out policy design during flexible and adaptive management while drawing on the experience.

Second, policy implementation. A policy can be deemed effective only when it is well implemented and generate good results. There are many factors that will affect the effectiveness of policy implementation in different countries. For example, the attitude of companies to policy measures is a key factor. Hence, an appropriate understanding of the basic conditions and scenarios for policy implementation is fundamental to ensuring that policies are well performed.

Third, institutional barrier which is a general factor which may cause the invalidity of the supporting policies. China and ASEAN have all set up environment departments, but the environment issue is under the management of many a departments. This kind of institutional design is bad for handling the environment issue.

Fourth, lack of resources. It refers to the personnel resources for policy drafting and implementation, facilities and equipment and capital. The environment departments of ASEAN countries encounter the shortage of all these resources.

On the new demand for policies, on one hand, it is necessary to handle new challenges. When time passes by, there will emerge new environmental issues and the prioritized areas will also change accordingly. Therefore, the policy shall also be adjusted, rather than just be restricted to specific time frame. The implementation of Sustainable Development Goals will provide opportunities for policy upgrade and better resource allocation to address the new challenge in a systematic way.

On the other hand, more scientific and feasible policies are needed, which shall include but not limited to environment institutional reform, stronger enforcement, promoting MBI and wider public participation. All of these shall be a critical part of establishing an effective environment policy framework.

## References

1. IEA.: Southeast Asia Energy Outlook 2013. IEA website (2013)
2. IEA.: Southeast Asia Energy Outlook 2015. IEA website (2015)
3. ASEAN Centre for Biodiversity.: ASEAN Biodiversity Outlook. Dolmar Press (2010)
4. Chazhong, G., Xu, K.: Innovative ecological compensation practice in Viet Nam: payments for forest environmental services. Environ. Prot. **16**, 70–71 (2010)
5. Wu, S., Yang, W., Zhao, Y., et. al.: Innovation and practice of aqua ecological compensation in Xin'anjiang river watershed. Environ. Prot. **42**(5) (2014).
6. Secretariat, A.: ASEAN Strategic Plan of Action on Water Resources Management. ASEAN Secretariat (2005)
7. Secretariat, A.: Fourth ASEAN state of the environment report. UNEP, ASEAN Secretariat (2009)
8. Resosudarmo, B.P.: Indonesia's clean air program. Bull. Indonesian Econ. Stud. **38**(3), 343–365 (2002)
9. UNEP.: Global Environment Outlook 4(GEO4): Environment for Development. United Nations Environment Programme. Progress Press Ltd, 2007.

10. UNEP.: Global Environmental Outlook 5 (GEO5)—Environment for the Future We Want. United Nations Environment Programme. Progress Press Ltd, (2012)
11. UNEP.: Global Environmental Outlook 6 (GEO6)—Asia-Pacific Outlook. United Nations Environment Programme. Progress Press Ltd, (2016)
12. Center, China-ASEAN Environmental Cooperation: ASEAN Environmental Review. China Environmental Science Press, Beijing (2011)

# Chapter 5
# China-ASEAN Green Development Practice

**Abstract** China and ASEAN have made significant progress in addressing environmental challenges. Industrial source pollution has been reduced, forest coverage and natural reserves (both number and size) increased, and ozone-depleting substances largely eliminated. The use of natural resources including water and energy resources has been to some extent decoupled from growing economic growth. The improvement of the overall environmental quality depends on the micro-level efforts; and such measures as nature reserve establishment and industry-specific environmental upgrades will help address environmental issues. Policies and measures that successfully generate promising achievements should be further maintained and rolled out. To this end, this chapter analyzes green development practices in China and the ASEAN Member States through case studies. The cases are selected according to the following criteria: making policy innovation, using cutting-edge technology/technologies, harmonizing socio-economic and environmental development, or showcasing local specialty and industrial representation. The issues faced both by China and ASEAN and cases on coordinated development of both sides are detailed before others. These cases (a) reflect the common features of similar environmental activities and/or solutions; (b) have the demonstration effects that enable the extension of technologies or concepts to other countries or cities with similar geographical characteristics; (c) are strongly applicable, referable, and acceptable in other areas; and (d) keep up with the times, with close linkage to social development and prevailing environmental concerns. In addition, the selected cases analyze the different development stages of different countries and highlight the time features of representative countries like Singapore, China, Viet Nam and Myanmar. They have distinct regional characteristics that reflect regional differences and commonality. These green development cases cover a wide range of fields, and demonstrate green development practices from the perspectives of circular economy, low carbon economy and green economy. They focus on the green practices in three major industries, in particular the application of circular economy approach in the primary industry (agriculture), secondary industry (construction and manufacturing) and tertiaryindustry (represented by tourism). The cases of resource utilization and climate change focus on emissions reduction and

 
China-ASEAN Environmental Cooperation Center, *China-ASEAN Environment Outlook 1 (CAEO-1)*, https://doi.org/10.1007/978-981-10-6211-7_5

water use, highlighting the importance of low carbon development. Air, water, soil and solid waste are considered in pollution prevention and control cases, of which joint regional air pollution prevention and control is favored in both China and ASEAN. Biodiversity conservation cases highlight the capabilities and contributions of nature reserves. Through problem-oriented analysis, this chapter provides a detailed description of green actions, including background, content, results, effects, influence, as well as representation and replicability. This chapter finds out that China-ASEAN green development practice features wide participation, scientific innovation, incentive mechanisms and region specific implementation. It also concludes that China and ASEAN shall take green development as a great opportunity to boost cooperation in technology transfer, policy innovation, resource efficiency improvement and cultural exchange, and jointly build the "maritime silk road".

## 5.1 Green Industrial Development

As developing or newly industrial countries, China and ASEAN Member States are embracing a new round of industrial and scientific revolution represented by circular economy and clean technology. The "cooperation, learning and sharing" between the two sides is conducive to addressing environmental challenges, and fostering a closer linkage that further strengthens cooperation in environment-friendly technologies and industries and joint policy studies.

China's green development must be accompanied by "four shifts": (a) from high pollution, high energy consumption, low added value to low pollution, low energy consumption and high added value; (b) from high-carbon economy to low-carbon economy; (c) from ecological aggression to natural harmony; and (d) from ecological deficit to ecological surplus. China's Circular Economy Promotion Law, which was put into effect on January 1, 2009, facilitated the reduction, reuse and recycling in the production, logistics and consumption process. Meanwhile, the green industry development fund is also needed to boost green technologies and industries by financing means and create new jobs while improving the economic structure. ASEAN is also challenged by environmental protection due to large economic dependence on natural resources. Industrial development should follow a model of sustainable development that entails lower energy and resource consumption and resource recycling and reuse in the production process. This section describes the green efforts of China and ASEAN in the fields of agriculture, manufacturing and construction, and the tertiary industry.

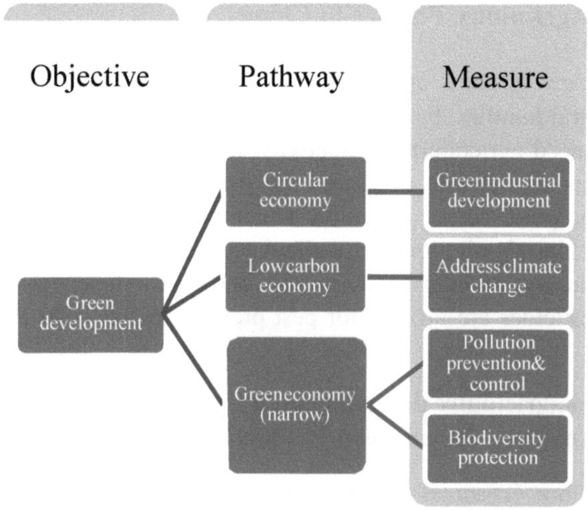


Fig. 5.1 Logic model of Chap. 5

## 5.1.1 Agriculture

Agricultural chemicals for farmland use lead to accumulation of harmful chemical substances in organisms through soil and water. Therefore, economic measures and legal means have been introduced in some countries to encourage and support pollution-free development and production of agricultural products. Agriculture is considered top of the ten priority areas of cooperation between China, a large agricultural country, and ASEAN where agriculture occupies an important status and role in the economy except Singapore and Brunei. Eco-agriculture will be an important direction of agricultural cooperation.

#### 5.1.1.1 Eco-Agriculture Practice in China

China guides new business agricultural entities to ecological agriculture. Among them, agricultural families, farmer cooperatives and leading agricultural enterprises represent advanced agricultural productivity at this stage. In practice, the application of high-quality manure serves as the major measure to effectively control agricultural non-point source pollution and improve agricultural ecological environment. Agricultural pollution is also alleviated as pesticides are reduced indirectly with the penetration of high-quality seeds, and high-efficiency, low-toxicity and low-residue pesticides are favored for pest prevention and control.

**Box 5.1 Eco-agriculture development in Eryuan County, Dali Prefecture, Yunnan Province—Broad Bean Production Base[1]**

Eryuan County organizes and mobilizes, through increased guidance and publicity, new agricultural business entities to actively develop eco-agriculture by establishing broad bean production bases.

- Demonstration area: 102,800 μ*
- High-quality broad bean seeds: 2000 tons
- Bio-organic fertilizer products: 800 tons
- Agricultural manure: 204,700 tons.

Comparison between demonstration area and non-demonstration area

- The average yield of fresh bean pod in the demonstration area exceeds 1200 kg/μ, and the weighted average yield in the county reaches 1229 kg/μ.
- The yield in non-demonstration area is 1092 kg/μ.

The yield of fresh bean pods is increased by 14.08 million kg in the demonstration area, indicating noticeable environmental and economic benefits.
Source: 1 μ = 0.067 ha.

Eco-agriculture requires improved agricultural quality and efficiency under available environmental conditions and improved plantation management through technical training. New agricultural business entities will continue to develop. In this context, a new-type agribusiness system with Chinese characteristic, which integrates a variety of collaborative and merging production and management organizations will take shape, pushing the transition from traditional agriculture to modern ecological agriculture.

[1]Source: http://www.ynagri.gov.cn.

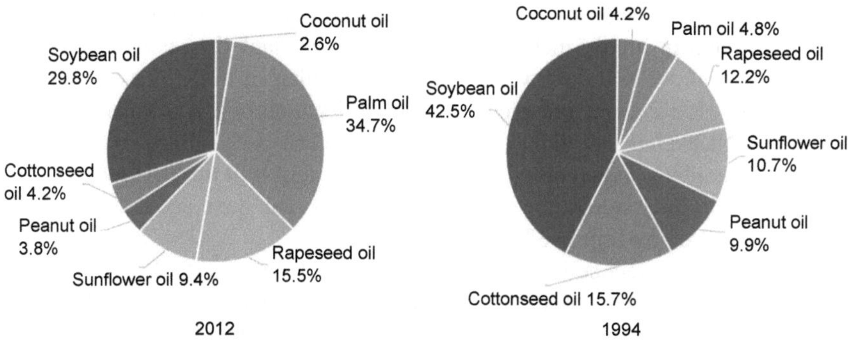


**Fig. 5.2** Global demand for vegetable oil

#### 5.1.1.2 Sustainable Palm Oil Production in ASEAN

The ASEAN Member States have long attached importance to the development of green agriculture. They are a large palm oil producer contributing 90% of the world's yield. The growing area of palm oil has expanded by 43% since 1990s onwards in ASEAN, mostly in Malaysia and Indonesia. The demand for palm oil increased the—fastest during 1994–2012. The yield of palm oil per unit of area is much higher than that of rapeseed oil, sunflower oil and soybean oil. Palm oil accounts for 34% of the world's edible vegetable oil while occupying only 5% of the growing area (see Fig. 5.2). However, the expansion of palm tree planting area has imposed huge pressure on the ecological environment. In 2004, the Roundtable on Sustainable Palm Oil Initiative (RSPO) was founded to promote the sustainable cultivation and use of oil palm. By reviewing all stages of the palm oil supply chain, RSPO encourages all RSPO members to reduce GHG emissions in the supply chain through voluntary mechanisms. RSPO sets a model for regional cooperation on environmental protection and rational resource use.[2]

#### 5.1.1.3 Eco-Village Project in Viet Nam

The Government of Viet Nam has carried out small-scale eco-village projects in the poor and ecologically fragile areas since 1990s. These projects are planned by the Ministry of Natural Resources and Environment and specifically managed by the Viet Nam Association of Ecological Economy and the Institute for Natural Resources and Environmental Strategic Policies. Project funding comes mainly from subsidies and aid from such countries as Sweden and France. Through these projects, rural areas are trained on strategies and methods for permanent and effective ecological rehabilitation and economic development. Fallow coastal hamlets in Quang Tri improved the drainage system in the rainy season and planted

[2]Source: http://www.rspo.org/en/home.

trees to prevent extension of desertification. Ba Vi and Ha Tay provinces created terrace farmland in highly inclined eroded zone and planted trees to prevent soil erosion and improve soil quality. Mangroves are replanted in the coastal wetlands in the Thanh Hoa Province in order to promote local shrimp reproduction. In the Y-Yen and Nam Dinh provinces, cultivating pond fish is combined with growing vegetable in gardens and comprehensive development of fruit trees. These measures promote ecological restoration, so that local residents can have access to more and better food and enjoy improved quality of life.

The eco-village practice in Viet Nam has many advantages: (a) wide scope of application. It can be carried out smoothly in developed and underdeveloped countries and regions and promote local economic development through ecological progress; (b) diversification of funding sources. The practice obtains grants from the Government and aids from other countries through the Government; (c) full use of local human and material resources which gives rise to new economic growth points and stimulates the enthusiasm of local residents; (d) recovery rather than creation, which is conducive to the protection of untapped resources and full and better use of tapped resources, thus contributing to sustainable development; (e) adaptation to the specific conditions, so that effective measures in line with local situation can be taken to produce substantial results; and (f) considerable economic benefits provide an incentive for local residents to engage in ecological restoration. The approach that favors full use of local resources is sustainable and worth extension. It should be noted that green action is expected to produce benefits, in order to maintain project vitality.[3]

#### 5.1.1.4 New Theory Agriculture (NTA) in Thailand

NTA is the guideline on land and water resource management with the aim of realizing the self-reliance of farmers and the harmonious co-existence of man and nature through land use master plan. It has been included by the Government of Thailand into the 8th Plan for National Economic and Social Development (1997–2001). NTA is aimed at progressive realization of sustainable development of agriculture at family, community and national levels by establishing large-scale agribusiness system and man-nature harmonious management system that supports the development of agricultural cooperatives and helps farmers become self-reliant based on rational planning of rural production and living space. Under the theoretical framework of NTA, families share the resources to reduce life cost and increase family income. With coordinated labor division among the community members, the community can realize self-reliance and have stronger inter-community communication so as to create external value chain and improve the financial guarantee for agricultural production. A rice-based diversified structure of agriculture has been established, which enhances food security and income security for farmers. The theoretic and practical studies on sustainable agriculture and

[3]Source: http://genoa.ecovillage.org/.

sustainable livelihood are also advanced. Practice shows that agricultural industrialization management and diversified business model under the NTA framework can effectively improve land productivity and land use sustainability and are worthy of promotion and study [1].

## 5.1.2 Manufacturing and Construction

### 5.1.2.1 China-Singapore Eco-Industrial Park

The major carrier of inter-governmental cooperative projects between China and Singapore is eco-industrial parks. Suzhou Industrial Park, which was established in 1994, had tremendous achievements not only in its exploration in industrial development, scientific progress, financial management, social governance, human resources development, and institutional innovation, but also in ecological and environmental protection and urban planning. Suzhou Industrial Park also features large ecological green lands, lakes and rivers. China-Singapore Tianjin Eco-industry Park further strengthened the concept of environment protection by saving and recycling resources including land, water and energy, treating and recycling wastes, and reducing the discharge of pollutants. It features low energy consumption, low pollution, low land occupation, high efficiency, and high quality service. It developed a urban green transportation system which is conducive to social justice, combined "passive" energy conservation with natural environment and "active" environment-friendly design with new technologies, adopted green standard for building's indoor design and built livable and environment friendly settings. It aims to build itself into a national demonstration zone for green development by 2020 and serve as the reference and a model for green development nationwide.

These two cooperative projects laid solid foundation for China and ASEAN to replicate feasible, replicable and promotable eco-industrial parks. At the same time, the cooperative projects between China and Singapore set up a very good model for regional reciprocal cooperation on response to climate change, environment protection, and resource and energy conservation.

### 5.1.2.2 Green Development of China's Automobile Industry

The automobile industry is an indispensable pillar of modernization. In 2014, the automobile industry contributed 1.44% of China's GDP, and ranked first in the world for seven consecutive years, in terms of the volume of production and sales. However, the entailed pollution has become a threat to human survival. In China, motor vehicle exhaust pollution and $PM_{2.5}$ becomes increasingly serious. In the face of environmental issues, the Chinese Government works together with enterprises to boost green development of the automobile industry.

**Box 5.2 Joint efforts of the Chinese Government and enterprises to boost green development of the automobile industry**

State Council:

- Issue the Opinions on Encouraging the Development of Energy-saving Vehicles with Low Emissions and other relevant policies and regulations.
- Practice higher motor vehicle emissions standards. China V and VI standards are getting stricter.
- Fuel standard is getting stricter. The fuel consumption limit for the first, second, and third stage of passenger vehicle is getting lower.
- Subsidy to clean vehicles.

Automobile companies:

- Establish R&D platform and result transformation base for environment-friendly vehicles.
- Independently design and introduce fuel cell vehicles and hybrid vehicles.
- Popularize new power technology to further improve fuel economy and reduce fuel consumption.
- Promote emission reduction technologies including waste gas recycling technologies for diesel engines and secondary supply technologies for gasoline engine.

The active development of energy-efficient and environment-friendly vehicles in China fits into the realities of energy supply and mass consumption, and helps protect the environment. A stable and continued strategic policy in favor of such vehicles and NEVs will encourage businesses to maintain high technical investment, clear and orderly direction of research and development, and steady improvement of technical capacity. Before achieving breakthroughs in highly efficient and reliable new energy technologies, the substantial progress in energy conservation and emissions reduction will boost green development of the automobile industry.

#### 5.1.2.3 Green Development of China's Construction Industry

The sustaining economic growth, accelerating urbanization process and improving living standards put higher requirements for the construction industry which used to follow the conventional mode of extensive development. Green buildings have attracted widespread concern with global climate change and low-carbon economic development. A complete set of energy-efficient and new energy technologies and intelligent technologies, including wall insulation, energy-saving windows and doors, heating control, and solar energy, have been rapidly used in buildings through pilot projects [2]. The economic benefits of green buildings are confirmed by the series of measures taken to green residential buildings and the extensive attention given to green office buildings.

**Box 5.3 Measures for green development of China's construction industry**

- Improve constantly the industrial chain, and incorporate conservation of energy, land, water and materials and protection of environment into different parts of the construction industry.
- Use economic levers and motivate the initiative of builders and users for green benefits.
- Develop incentives specific to national conditions at national and local levels, e.g. ecological compensation and tax reduction and exemption for green buildings to balance the investment interests of developers, users and the government.
- Promote standardization system and improve the green certification system of construction industry.
- Enhance the quality of personnel, and educate the participants of the construction projects with green development concepts.
- Foster green building consumption ideas with industrialization as the means, and improve the awareness of the consumers on green buildings.

Towards the goal of green buildings, China's construction industry should actively develop and apply technologies for energy efficiency, environmental protection, and resource recycling. An important way is to industrialize production, with standardized construction components and integrated construction process. Specific to residential buildings, it is necessary to establish new development ideas, develop core technologies for saving energy, land, water, and materials, and build an industrial chain covering the whole life cycle of buildings.

#### 5.1.2.4 Sustainable Consumption and Production Policies and Practices in Malaysia

Malaysia plans to become one of the most sustainable and inclusive high-income countries by 2020 through a series of reforms. The patterns of sustainable consumption and production (SCP) will be strengthened as the unsustainable behavior has severely undermined environmental sustainability. Committed to an integration of economic development, environmental protection, and social inclusion, SCP stresses the achievement of sustainable development by practicing green consumption and reducing the cost of production cycle. The Government of Malaysia strengthens policy implementation and develops institutional framework in order to achieve SCP based economic development. In addition, promoting sustainable consumption and production is listed into the 11th Malaysia Plan, in which the plans implemented are specified as follows:

**Box 5.4 SCP plans implemented in Malaysia**

- Green Building Index (GBI): designed to foster sustainability of the construction industry and its relationship with the environment and raise practitioners' awareness of environmental protection;
- Green product directory (MyHijau): a comprehensive guide to a wide range of green products, services, and systems for consumers' reference;
- Natural Environment Partnership: intended to increase environmental awareness of local communities in parliamentary constituency;
- Low Carbon Cities Framework (LCCF): a national rating system for environmental impact assessment based on the total carbon emissions;
- Green procurement: The share of government green procurement has taken 12–15% of GDP. It has become an important means to stimulate economy and promote supply.

Over the past decades, the number of residents and local communities that join in the SCP campaign in Malaysia has been on the rise. Green technologies have been widely used and remarkable progress achieved in announced awareness and commitments to environmental protection. The involvement of a variety of stakeholders greatly intensifies the efforts to implement the policy and provides strong support to communities for the SCP campaign. The private sector has focused efforts on capacity building for providing green products and services and boosts exports to high-demand European, US and Japanese markets. In addition, Malaysia has acquired a wealth of experience in the SCP policy implementation through policy studies and pilot projects, which provides valuable reference for other countries.[4]

#### 5.1.2.5 Thailand's Sustainable Consumption and Production Strategy

Thailand launched a sustainable consumption and production strategy to facilitate the implementation of its 10th National Economic and Social Development Plan. The Government of Thailand also set up the Competitiveness and Sustainable Development Promotion Committee to formulate and implement the government green procurement policy which requires the government to purchase products with less negative impact on the environment in order to encourage production of such green products. Aimed at balance between present and future generations in social satisfaction, food security and self-sufficiency, the sustainable consumption and

[4]Source: OIDA International Journal of Sustainable Development, Vol. 08, pp. 23–32, 2015.

production strategy has been successfully implemented, according to preliminary indicators. Producer-related indicators show reduced energy intensity; consumer-related indicators show more landfill recycling, which indirectly pushes up the sustainable development indicators of the country; and the overall indicators imply significantly reduced ecological footprint.[5]

**Box 5.5 Thailand's specific measures for sustainable consumption and production strategy**

- Disseminate industries of profitable cleaner production and successful financial models of self-sufficient families and communities.
- Seek government support for green products and services.
- Establish market mechanisms for recycled and reused products.
- Rebuild the economic structure in order to increase the proportion of service industry.
- Promote the establishment of common values and strategic vision of sustainable consumption and production.
- Enhance business ethics and effective management.
- Encourage public participation in the formulation, implementation and evaluation of sustainable consumption and production policies.
- Boost exports of green products.

The Siam Cement Public Company Limited (SCG) has been committed to green business based on good corporate governance and sustainable development. To this end, it develops two strategies. Regarding green production, in line with the "reduction, reuse and recycle" (3R) principle, SCG uses environment-friendly technologies to improve machine efficiency to ensure more efficient energy use, and also carries out research on alternative fuels. Regarding green products, SCG sticks to the development and research of environment-friendly products and strict application of eco-labels. Under the guidance of "green business", the life cycle assessment of products is introduced to ensure effective use of resources and promote the use of renewable resources. Owning to eco-label and environment-friendly products, SCG creates a brand image favorable for environmental protection and sets an excellent model for energy conservation and emissions reduction in the country (see Table 5.1). Many other companies in Thailand began to follow suit, in order to improve business image and environmental control capacity.[6]

Government's practice and promotion of SCP will effectively mobilize the general public and the private sector. As a major player to boost green development, the government can use a policy portfolio made up of visibility and awareness campaigns, incentives, and market mechanisms. Incentives can play a dominant

[5]Source: www.scp-thailand.info/.

[6]Source: UNEP, http://www.scg.co.th.

**Table 5.1** SCG's achievements in energy conservation and emissions reduction

| Environmental performance | 2009 | 2010 | 2011 | 2012 | 2013 |
|---|---|---|---|---|---|
| Total energy consumption/GJ | 135.49 | 140.68 | 167.68 | 174.58 | 175.93 |
| Heat consumption/GJ | 123.77 | 128.30 | 153.65 | 159.70 | 161.50 |
| New energy consumption/GJ | | | | | |
| Biomass | 8.84 | 7.31 | 9.10 | 9.86 | 10.26 |
| Amount of industrial waste | 10.90 | 10.71 | 8.78 | 8.35 | 10.70 |
| Electricity consumption/million kW h | 3254 | 3441 | 3895 | 4133 | 4008 |
| Water consumption/million $m^3$ | 91.89 | 95.50 | 104.55 | 110.80 | 104.52 |
| Circulating water ratio/% | 10.78 | 9.99 | 9.90 | 9.52 | 6.98 |
| GHG emissions/million tons | 18.92 | 19.66 | 20.73 | 21.96 | 21.51 |
| Hazardous waste/kt | 13.44 | 17.02 | 12.21 | 14.65 | 19.60 |
| Non-hazardous waste/kt | 653.60 | 1176.12 | 1305.30 | 1215.07 | 1130.67 |

role at the early stage, and market mechanisms at the latter stage, which will help sustain green development.

#### 5.1.2.6 Cleaner Production in Industry (CPI) of Viet Nam

In order to address the root causes of industrial pollution, the Government of Viet Nam has formulated and implemented laws related to environmental protection and launched the CPI scheme that guides and supports enterprises to improve efficiency of resource use and prevent the generation of pollutants (see Table 5.2). CPI aims to improve the living conditions of workers and communities and strengthen capacity building of producers by reducing industrial pollution and increasing productivity. It will provide practical experience and demonstration on cleaner production that promotes sustainable industrial development. The scheme includes: (a) industrial enterprises that adopt cleaner production and accept capacity-building assistance commitment; (b) specific target provinces; and (c) effective, strategic, and replicable cleaner production technologies. Totally eight activities have been carried out under the scheme, as shown in the table below:

The CPI scheme was implemented from September 2005 to December 2011, It funded demonstration projects in 61 enterprises distributed in 5 provinces, including Thai Nguyen, Phu Tho, Nghe An, Quang Nam and Ben Tre. The CPI projects have generated pragmatic economic benefits and significantly improved the compliance of Viet Namese industries. These practices can be extended and replicated in enterprises of different sizes in different industries, especially SMEs.[7]

[7]Source: Viet Nam: Some Good Sustainable Development Practices, Hanoi, May 2012.

**Table 5.2** CPI activities

| No. | Activity | Content |
|---|---|---|
| 1. | Develop a national strategy | Identify industrial cleaner production as a national strategy covering large, medium and small-sized industrial enterprises |
| 2. | Establish monitoring system | Establish an operations department in the Ministry of Trade and Industry, to promote and monitor the implementation of cleaner production in enterprises |
| 3. | Strengthen weak links | Develop appropriate action plans to make up for the existing promotion system and regulatory framework |
| 4. | Divide functional departments | Establish a subordinate functional department in the Ministry of Trade and Industry, responsible for supporting SMEs to implement cleaner production |
| 5. | Improve the management system | Include cleaner production into the provincial pollution management plans |
| 6. | Implement pilot projects | Carry out 40 demonstration projects in the priority areas and regions |
| 7. | Balance benefits | Promote production in compliance with environmental, health and safety regulations while improving economic benefits of demonstration enterprises |
| 8. | Sum up lessons learned | Sum up experience of demonstration projects for extension in the relevant sectors and regions |

#### 5.1.2.7 Asia Pacific Resources International Holdings Ltd in Indonesia

Asia Pacific Resources International Holdings Ltd (APRIL) is an Indonesian developer of fiber plantations situated mostly in ecologically fragile tropical peatlands, and its long-term development rests on the sustainable development of peatlands. The peat layer plays an important role in keeping water sources and serves as a high-density "carbon sink" through deforestation, droughts and fires. Hence, the sustainable use of peat resources and prevention of massive carbon emissions and fires require adequate moisture content of the peat layer.

In 2007, APRIL launched a three-year science-based management support project (SBMSP) to raise public understanding of hydrology, ecology, and matters relating to the sustainable management of peatland. Regarding energy efficiency and emissions reduction, paper mills follow best global practices for environmental protection, including monitoring the emissions of waste gas, waste water and solid waste on a regular basis. Investment in new technologies is boosted to achieve the overall compliance with emissions standards ($NO_x$, $SO_x$, TRS, and $CO_2$). Efficient use of all raw materials is the guiding principle of factories, and all pulp and chemical residues in paper mills are recycled and reused. At the same time, APRIL reduces the use of fossil fuels and relies on byproducts of the production process to satisfy its 75% energy demand. A scientific and technical support team is set up to support APRIL's practices.[8]

The responsible management practices of APRIL, including best practices in protection of high conservation value forests (HCVFs), fully reflect the incorporation of environmental protection into the production process to achieve sustainable production and double harvests of economic benefits and environmental benefits. APRIL successfully establishes a sense of responsibility in business sector to encourage wise use of natural resource in making products favorable for higher quality of life.

## *5.1.3 Tertiary Industry*

### 5.1.3.1 Mangrove Conservation and Tourism in China and ASEAN

Mangroves are special phytocoenosium in tropical and subtropical coastal areas and largely distributed in hidden gulf and estuaries. China's mangrove forests total 15,122 ha. With greater awareness of the value of mangroves and bigger pressure on resource protection as well as the demand for eco-tourism, China established 5 national mangrove nature reserves and 16 provincial reserves and developed eco-tourism. ASEAN Member States, located in tropical and subtropical areas, also enjoy rich mangrove resources. ASEAN Member States like the Philippines and Malaysia are also developing their mangrove tourism.

Dongzhaigang Mangrove Nature Reserve in China's Hainan province is the first reserve of this kind in China. After its establishment in 1987, it developed mangrove tourism and had local villagers involved in the management which helped to increase their income and decreased the destructive development of mangroves. Guangxi Shankou Mangrove Nature Reserve opened land and cruise tourism in 1992 and put forward eco-tourism design ideas in 1997 including small, scattered and hidden natural trails, exhibition halls, science tourism, river bank tourism, maritime tourism and memorial planting, etc. In 2002, it established a mangrove science center and species display platform, held mangrove planting volunteer activity and developed mangrove posters and post cards under "Shankou Mangrove Nature Reserve Ecological Tourism Development and Public Education Project" supported by UNESCO. Shenzhen Futian Mangrove Nature Reserve was established in 1988. As an important part of Shenzhen gulf wetland, the reserve protects subtropical bay mangrove swamps and established such sites as Futian mangroves and bird nature reserve science center, mangroves park, birds park, South China Botanical Garden, art corner, green corridor, fishing area and exhibition hall for nature protection, integrating the functions of science education, coastal culture, natural landscape and leisure sightseeing.

---

[8]Source: http://www.aprilasia.com.

Bohol Province of the Philippines has 5 mangroves nature reserves. After integrating other resources, local government and NGOs offered tourism products for both independent travel and team tour. For independent travel, kayaking and raft are available for tourists; for team tour, professional guides will introduce mangrove knowledge and the visitors will have a chance to go to the mud flat. After the trip, each tourist will be given a mangrove sapling to plant on specific mud flat to encourage them to protect mangrove.

Langkawi Mangrove Reserve is located in Kota Kinabalu Wetlands in Sabah, Malaysia. It is used primarily as a model wetland center for the purpose of conservation, education, recreation, tourism and research and was selected as a world heritage site in 2007. Langakawi Mangrove Reserve developed tourism using the cliff painting of Gua Cerita Cave of 4000 years history, the quiet Jurassic island and the natural salt lakes. The management authorities also set up artificial habitat for bats in the limestone cave of Kilim Geoforest Park. Tourist activities include round tour of the mangroves, feeding eagles, and special landscape sightseeing.

The eco-tourism programs of the mangrove nature reserves in China and ASEAN have all integrated protection, tourism and education together. Such programs created a triple win situation for environment, economy and society. As a result, mangrove ecosystems are protected, local income and livelihood improved, and public awareness of and passion for environmental protection enhanced.

#### 5.1.3.2 Green Development of Tourism in China

An increased share of environment-friendly tertiary industry, especially the service industry, is fundamental for green economic transformation and applicable to sustainable development strategies of developing countries as an important factor for transition to developed countries. In recent years, tourism has boomed in the tertiary industry. Recognizing the potential negative impact of tourism development on the environment, China has established tourist ethics and industry ethics standards in social and economic systems. Taking advantage of science and technology, green tourism facilities are built, covering garbage classification and recycling; green and renewable energy sources are used to conserve energy and resources and reduce $CO_2$ emissions; water conservation and water pollution control are incorporated in their daily operations; green food and green travel are also favored.

**Box 5.6 Development of eco-tourism economy**

Relying on eco-agriculture, holiday eco-villages are practicing green tourism guided by the business philosophy of "green, eco-friendly, and sustainable development". The measures taken include:

- Reduce fossil energy consumption;
- Recycle water;
- Recycle materials:

- Brew with grains and feed livestock with distillers grains;
- Produce biogas from excrement of tourists and livestock for cooking and lighting;
- Apply biogas residues and wastes into farmland as base fertilizer; and
- Introduce crabs to paddy fields to repel insects, eat weeds, and loosen the soil to avoid use of pesticides and fertilizers in the planting process.

The Chinese tourism sector promotes conservation-oriented industrial development by building a green industrial system in line with green development requirements. It advocates green consumption in tourism and guide tourists to strengthen the awareness of environmental protection and tourism businesses to follow the ecological law. Good results are achieved through the organic linkage of stakeholders in tourism development, consumption and environmental protection. A green tourism based environmental protection system has been built, with the development of tourism goods and services and management standards in accordance with local characteristics and green development requirements.

#### 5.1.3.3 Eco-Tourism in ASEAN

Eco-tourism has achieved rapid development in Thailand owning to support from governments at all levels. With transfer of tourism management functions to local governments, local government managers update their knowledge, including tourism planning and management standards, and make provisions for long-term protection of tourist attractions. Tourism management and research institutions develop scientific plans that encompass correct and comprehensive development strategies and objectives and appropriate policy and regulatory constraints, to avoid blind and unlimited development of eco-tourism. The Government defines the scope of nature reserves and boundaries of restricted development areas. In addition to such scientific planning, laws and regulations provide an important guarantee for sustainable development of tourism, such as the Tourism Act, Enhancement and Conservation of National Environmental Quality Act, Hotel Act, and Tourism and Tourist Guide Business Act. With dissemination and education on sustainable tourism, the idea of environmental protection and sustainable development has been greatly enhanced in the minds of the people [3]. In addition, the development of Internet and information technologies has raised tourists' awareness of environmental protection, thereby effectively promoting sustainable tourism.

RumahTuri in Indonesia[9]

Rumah Turi, Indonesia's first eco-friendly hotel, represents a preferable combination of rich natural culture and solo art by using green design technologies. It successfully uses renewable materials by renovating old houses according to local conditions and reutilizing almost all buildings materials available. Wall coverings with natural clay color reduce the use of paint and improve indoor air quality. Hotel furniture is mostly made of recycled wood, and 0.2 W LED lamps cut 30–40% electricity consumption. The architectural design allows configuration of structures in harmony with natural landscape, and ensures green and eco-friendly process of construction, presenting agreeable commercial and educational functions.

El Nido Resorts is a group of ecological resorts in the El Nido and Taytay municipalities in Palawan, Philippines. It adheres to the principle of sustainable development in highly responsible management of this privileged natural environment and stresses the importance of ecological protection to all visitors. Visitors can enjoy a variety of wonderful activities without undermining the environment, such as diving, cannoning and kayaking, and hiking. The daily operations of the resorts always reflect the idea of sustainable development. By providing job opportunities, the resorts offers stable income and skills training to local residents. While procuring 60% of the vegetables from local organic farms and 90% of chicken from local village backyards, the resorts operates its own organic farm that hires local farmers and trains them the right cultivation and management methods. El Nido Resorts has won a lot of international awards and been committed to

[9]Source: http://sindicato-uno.cl/comunicados/skins/intense/rumah-turi.

formation of a sustainable, benign and mutually beneficial business model that drives local economic development.

**Box 5.7 Sustainable operational measures of El Nido Resorts in the Philippines**

- Recycle materials;
- Reduce disposable containers;
- Protect water resources;
- Improve energy efficiency and make full use of the island's abundant sunlight resources; and
- Provide low-impact customer experience.

Inle Lake is located in the central highlands of Shan State, Myanmar, where the renowned tourist destination Phaung Daw Oo Pagoda attracted more than 150,000 foreign tourists in 2013. The Inthar build houses with tall stakes in lakeside or shallow waters, creating floating villages and floating gardens which represent unique local culture. Floating gardens were introduced in the early 1960s, made of the mass of grasses, reeds, sedges, and other aquatic vegetation and used to grow tomatoes, flowers, vegetables, and sugarcane. Under this hydroponics system, despite of the benefits, uncontrolled agrochemical use has caused water pollution of the lakes. Other threats include shrinking water area, eutrophication, and local fish species threatened with extinction.

In response, the Government of Myanmar has developed an effective protection mechanism by setting up a Working Committee for Protection of the Inle Lake. Joint cooperative departments in charge of irrigation, forest and agricultural services have identified the respective work as follows: controlling the enlargement of village and floating gardens, protecting the existing forests and trees planting in the catchment, and maintaining water resources for hydropower stations. The National Committee for Sustainable Development and Environmental Protection of the Inle Lake also developed a five-year protection and restoration plan and the forestry department of Shan State launched the Inle Lake greening project (2010–2015). The President of Myanmar witnessed the establishment of Inle Lake Conservation Agency in 2014 and the Government of Myanmar committed an annual investment of 100 million kyat (100,000 US dollars) in the protection of the Inle Lake. Projects for Inle Lake protection have also been supported by international funds.

Borneo, politically divided among Indonesia, Malaysia and Brunei, is the third largest island in the world after Greenland and Papua New Guinea. It is known for pristine tropical rainforest and numerous rare and endangered species of wild fauna and flora. Local tourism companies try to bring benefits to local residents from every foreign tour project while maintaining operations in full compliance with the principles of sustainable development of the economy, society and environment.

Part of tourism revenue flows into non-profit funds for reforestation and local rainforest restoration.

Which is usually completed by the local non-profit organizations. In 2007, eight ringgit out of the money spent by each visitor was used to plant trees in order to alleviate the greenhouse effect, and a larger share was paid to similar projects.[10]

Many eco-tourism projects in ASEAN well reflect the necessity and rationality that companies assume their social responsibility and the consequent enormous economic and environmental benefits. Sustainable tourism requires tourism development in harmony with the nature, culture and human environment and mainly serves for the national and local economies and higher living standards. The sustainable development of tourism drives economic development which enables better protection of the environment and cultural resources, and a favorable ecological and cultural environment can also support sustainable tourism. It is the positive interaction between tourism and economic development that ensures tourism achieve true sustainable development.

## 5.2 Resource Utilization and Response to Climate Change

Climate change is a global issue. The global average surface temperature has risen by 0.74 °C over the 100 years and the global land precipitation increased by about 2% since 1970. In mid- and high-latitude continent in Northern Hemisphere, the precipitation has increased significantly, but the spatial-temporal distribution

[10]Source: http://www.elnidoresorts.com/, http://www.borneoecotours.com/.

**Table 5.3** Dreadful disasters in China and ASEAN (2008–2015)

| Time | Event |
|---|---|
| May 2008 | Strong tropical storm hit Myanmar |
| July 2011 | A once-in-60-years flood took place in Thailand |
| July 2012 | Beijing suffered 61-year strongest rainstorm and flood disaster |
| June 2013 | Flood and hailstorm hit Jianghuai and Jianghan region, Sichuan basin |
| November 2013 | Super Typhoon Yolanda hit the Philippines |
| July 2014 | Ramasoon super typhoon |
| May 2015 | Flood and hailstorm hit Jiangxi and Fujian provinces |

remains uneven. In recent years, much attention has been given to the relationship between climate change and natural disasters. Meanwhile, China and ASEAN Member States have suffered frequent occurrence of natural disasters in recent years, exerting huge impact on the society and economy of different countries and even people's lives (see Table 5.3). Given this, China-ASEAN capacity building for responding to climate change through enhanced regional cooperation and exchanges becomes increasingly important.

China and ASEAN are facing the serious challenge of climate change. China's Policies and Actions for Addressing Climate Change (2011) and the White Paper on Overall National Strategy for Adapting to Climate Change have markedly improved the capability and level of China's response to climate change. For example, in June 2014, China-ASEAN International Seminar on Climate Change was held in Kunming, China. The seminars between China and ASEAN in recent years also play a positive role in building up the regional capacity of responding to climate change. This section focuses on green actions that China and ASEAN have taken to deal with climate change, covering water resources use, forest carbon sinks, waste recycling, and biomass energy.

### 5.2.1 *Improved Utilization of Water Resources and Oasis Creation in Xinjiang, China*

In the context of climate change, Xinjiang plans ahead for development and utilization of water resources by strengthening management and building key infrastructures. A positive position in use of water resources, secured by accelerating construction of key infrastructures for water allocation, and fostering early the use and development system in water-abundant areas, can ensure the long-term substantive solution to the shortage of water resources in the eastern border areas. Aksu River, a key source river of the Tarim River, plays an important role in the river ecological and environmental protection and has 65% or so of the average annual runoff from Kyrgyzstan. Xinjiang has actively expedited substantive cooperation with Kyrgyzstan in scientific research, hydropower development, and construction, in order to protect the ecological environment of the Tarim River.

**Box 5.8 Measures taken in Xinjiang of China for adaptation to climate change**

- Implement strict water resources management;
- Control total water use and irrigation areas;
- Strengthen orderly exploitation and monitoring of groundwater;
- Conduct special studies on water cycle changes in key river basins and strategies for adaptation to climate change;
- Improve development of strengthen contingency plans for natural disasters and water resources reserve system safety program; and
- Increase water supply for such water-scarce northern and eastern areas and further tap the potential of development and utilization in water-abundant areas.

Considering water scarcity and rich land reserves, comprehensive studies on the resources, environment and economic development in the oasis region, as well as climate change are carried out, laying a strong foundation for ecological environmental protection. Based on these studies, people can prevent the expansion of desertification and secondary salinization of soil, but also gradually develop efficient eco-agriculture. The capacity for sustainably utilize water resources has been significantly improved through adjustment to water supply and demand structure, and basically meets the rational water demand for economic and social development. Marked improvement is seen in not only water use efficiency and effectiveness, but also the water environment [4]. The overall direction of future development and utilization of water resources in Xinjiang is to build a water-efficient society in an all-round way by way of industrial restructuring, and achieve sustainable use of water resources through science-based development and management.[11]

## 5.2.2 *Forest Carbon Sinks and Bamboo Forestation*

By protecting forests in biodiversity hot spots and biodiversity-rich wilderness areas, Conservation International, an environmental organization, helps China cut GHG emissions by 1–2 billion tons annually,[12] which is approximately equivalent to 3–5% of the targeted global emission reductions by 2030. Like other forests, bamboo forests can absorb $CO_2$ and achieve carbon sinks. However, a process-wide

[11]Source: Urgency for Sustainable Use of Water Resources Protection in Xinjiang in the Context of Climate Change.

[12]Source: http://www.conservation.org.cn.

methodological guidance on bamboo forest carbon sinks remains absent as the calculating and monitoring methods are not clarified. Prior to 2012, dozens of methodologies for afforestation and reforestation CDM projects were approved, but none was applied to bamboo forestation in the international standards for verified emission reductions. As a result, the contribution of bamboo forests to $CO_2$ absorption and emissions reduction cannot be recognized by the market, making it difficult to enter the emissions trading market. The situation did not change until 2012 when The Nature Conservancy (TNC) introduced the "Panda Standard"—methodology for forestation using species including bamboo.

**Box 5.9 Panda Standard: methodology for forestation using bamboo and non-bamboo trees**[13]

In 2012, the Panda Standard was formally approved and named "Panda Standard—Methodology for Forestation of Degraded Land Using Bamboo and Non-bamboo Trees", with TNC's success of experimental bamboo forest project in Nannuo Mountain. The world's first methodology for bamboo forestation is of very significance:

- Clarify the direction of quantitative standard for the entry of the bamboo industry into the carbon market;
- Lay the methodological foundation for application for Panda Standard bamboo forestation projects recovery of degraded land, and sequestrated carbon estimates;
- Provide process-wide methodological guidance on China's development of distinctive forestry carbon sequestration projects; and
- Offer reference for forestry carbon sequestration projects in other developing countries.

The forest carbon sequestration projects of Conservation International helps China take an international leading position in the conservation of natural resources to tackle climate change. At the same time, the Panda Standard as a methodological outcome can be applied to national dominant plants in developing countries while drawing on project experience in China, in order to promote the cultivation of native plants and thereby play a role in carbon sequestration. In the process of mitigation and adaptation to climate change, Conservation International promotes the introduction of national policies to encourage market-based economic solutions to protect forests and reduce emissions, including pioneering investment in public and private funds, and also helps develop market trading standards for economic benefits of forests.

[13]Source: http://www.tnc.org.cn.

### 5.2.3 NEWater in Singapore

Singapore produces drinking water by reclaiming and treating sewage to address the problem of water scarcity. When sewage flows through the drainage to sewage treatment plants, the solids and certain bacteria are removed, and several other barriers are designed in order to make the reclaimed water potable: (a) microfiltration through membrane 100 times thinner than human hair to remove bacteria; (b) reverse osmosis (RO) through semipermeable membrane to remove salts, virus and drugs; and (c) disinfection using high-intensity ultraviolet and hydrogen peroxide to inactivate all organisms. Singapore claims that the reclaimed water is almost as pure as distilled water and used in a large number of industries, including semiconductor and air conditioning, with gradual extension to the civilian sector. Drinking water that is made from toilet water may be unacceptable to many people, but recognized by about five million residents in Singapore.[14]

Recycled wastewater has been widely used for agricultural and industrial production, but for drinking water supply in only a few brave countries, such as Singapore. NEWater will provide useful experience for cities in developed and developing countries, specific to the formulation of policies on water use reduction and alternative water sources for household and industrial water supply. If widely accepted by the public, this technology will help address the global water shortage.

### 5.2.4 Barangay (Commune) in the Philippines

The Philippines Republic Act No. 9003 provides Barangay as the partner for environmental protection, especially for solid waste management. The environmental stewardship in Barangay uses carts or bikes with sidecars to collect biodegradable and renewable wastes which are then supplied to the material regeneration agencies for classified treatment. The work greatly reduces waste destined for landfill and helps the government cut billions of dollars of waste collection and treatment costs. Moreover, the sale of recyclable materials brings additional income for local communities. The partnership produces huge health benefits by creating a favorable environment, while cutting the government expenditure on environmental protection.

Adapted to local conditions, this approach fully utilizes grass-roots power and resources at hand and presents satisfactory effects. In China, many of the numerous cyclists support the cause of environmental protection, posing great potential for solid waste collection and disposal. If mobilized through advocacy and

[14]Source: http://www.pub.gov.sg/water/newater/Pages/default.aspx.

demonstration, they are believed to play a great role in reducing solid waste treatment cost.[15]

### 5.2.5 Biomass Application in Thailand

Thailand's primary energy consumption is over-reliant on natural gas. Against the world backdrop of increasing the share of clean energy, Thailand does the opposite, pushing up the share of natural gas in power structure to 70%. To diversify the energy sources, the Government of Thailand develops biomass energy. Thailand is the largest biofuel producer in Southeast Asia and the third largest palm oil producer in the world. Biomass is an important renewable energy and the second largest energy source in Thailand. It is widely used in home cooking and heating, manufacturing and traditional industries.

Sugar cane, rice, oil palm and wood waste are the four major sources of biomass in Thailand, which has an annual production of up to 80 million tons. There are also many potential biomass sources. Currently, agriculture provides a lot of biomass feedstock with large-scale use of rice mills and sugar machine, including rice bran, bagasse, fiber and shell. Another important source of biomass feedstock is the timber industry. The major biomass technologies include gasification, combustion, pyrolysis, and biogas, and existing biomass applications include biomass power plants, liquid fuel, biogas production, and improved stoves and furnaces. Thailand's coconut biomass combustion technology is a clean technique that converts waste into energy power. Relying on this technology, under the Mahachai Green Power project, the Samut Sakhon Province established a 9.5 MW power plant that uses coconut waste to generate power. The design realizes efficient use of fuels or mixture of a variety of fuels, and introduces flue gas cleaning system to reduce emissions to meet the standards. Ash after combustion will be used as farm fertilizer or filling materials for buildings. This power plant is networked to the grid in 2015 and entitled to tax subsidies for biomass energy provided by the Government. From the Government's point of view, the use of local waste materials to generate power can produce significant environmental and social benefits.[16]

Through a number of initiatives, Thailand utilizes local biomass resources to convert waste into treasure, which improves product added value, and also makes full use of post-production waste. Meanwhile, tax subsidies for biomass development are combined with financial management ideas to promote green technologies. Joint ventures with upstream and downstream industries are built to generate economies of scale.

---

[15]Source: Environmental Management Bureau, Department of Environment and Natural Resources, Philippines.

[16]Source: http://www.dpcleantech.com/.

### 5.2.6 Promoting Energy Conservation in Small and Medium Scale Enterprises in Viet Nam

Saving energy and improving energy efficiency is the urgent challenge faced by SMEs in Viet Nam. This is not only economically and technically difficult, but also ecologically challenging as it involves carbon dioxide emissions reduction and environmental protection. To this end, the Viet Namese Ministry of Science and Technology launched the Project of Promoting Energy Conservation in Small and Medium Scale Enterprises (PECSME), with an aim to enhance the overall energy efficiency of enterprises and reduce GHG emissions through promotion, digestion and absorption of energy-efficient technologies and management, maintenance and operational methods.

PECSME encompasses six components, namely policy and institutional support, communication and awareness, technical capacity development, energy efficiency services provision support, financial support, and energy conservation demonstration. Five years after launched, PECSME has become one of the most successful projects in the cooperation between the Viet Namese Government and the United Nations Development Programme (UNDP) in the field of sustainable development and energy efficiency. It has created a favorable environment for technological innovation and digestion and absorption of modern energy efficiency management methods, so that SMEs can achieve economic benefits, technological innovation and environmental benefits, thus making a contribution to environmental pollution mitigation in five industries. As of June 2011, a total of 543 energy efficiency projects involving ceramics, brick, paper, textile and food industries were completed; 25 provinces expressed support for energy efficiency improvement and technological innovation;

and approximately 25 suppliers of energy efficiency services and technologies offered technical support for over 500 SMEs. With the implementation of the project, the costs of production were cut by 10–50% and productivity increased by 30%, and product quality and corporate competitiveness have been significantly improved. A noticeable reduction is also seen in the environmental resistance to industrial development. As far as SMEs are concerned, the oil consumption was reduced by 232,000 tons, $CO_2$ emissions by 944,000 tons, and energy costs by 24.3%.

Through the provision of economic and financial support, the Government of Viet Nam successfully promotes and applies energy saving and environmental protection technologies among SMEs in five industries for the purpose of reducing production costs, increasing productivity, and improving product quality and competitiveness. It represents a fruitful practice in SMEs sustainable development, but also provides an important reference for industries of different sizes.[17]

### 5.2.7 *Achievement of Climate Change Emission Reduction Targets in Indonesia*

Indonesia is vulnerable to the impact of climate change and recognizes the mitigation of climate change as a major challenge. According to the Bali Action Plan adopted at the 13th Conference of the Parties to UNFCCC and the outcomes of the 15th and 16th Conference of the Parties at Copenhagen and Cancun, the Government of Indonesia has committed at the G20 Pittsburgh Summit to, by 2020, cutting the national carbon emissions by 26%, or even up to 41% if international support is forthcoming. In 2011, the Government has released a national action plan reflecting the abatement targets (see Table 5.4).

Indonesia has included climate change into National Development Plans and the Medium-term Development Plan (2010–2014). Currently, land (especially forests and peatlands) use and related changes account for 87% of nationally determined emission reductions. To achieve the abatement targets, the No. 6 Presidential Decree of 2013 requires suspension of the issuance of new license to primary forests and peatlands, in order to prevent forest destruction and peatland conversion, promote sustainable forest development, and strengthen peatland management. The National Action Plan, implemented by the Ministry of Agriculture and Forestry, mainstreams climate change into the development of all industries, making low-carbon economy an important requirement for mitigating climate change and reducing poverty.

[17]Source: Vietnam: Some Good Sustainable Development Practices, Hanoi, May 2012.

**Table 5.4** Indonesia's national action plan for reducing GHG emissions

| Sector | Emissions reduction plan (one billion tons $CO_2$) | | | | | |
|---|---|---|---|---|---|---|
| | 26% | Percentage | 41% | Percentage | Total | Percentage |
| Forests and peatlands | 0.672 | 87.6 | 0.367 | 87.0 | 1.039 | 87.4 |
| Waste | 0.048 | 6.3 | 0.030 | 7.1 | 0.078 | 6.6 |
| Agriculture | 0.008 | 1.0 | 0.003 | 0.79 | 0.011 | 0.9 |
| Industry | 0.001 | 0.1 | 0.004 | 0.9 | 0.005 | 0.4 |
| Energy and transportation | 0.038 | 5.0 | 0.018 | 4.3 | 0.056 | 4.7 |
| Total | 0.767 | 100.0 | 0.422 | 100.0 | 1.189 | 100.0 |

### 5.2.8 *Pollution Prevention and Control*

In the process of advancing ecological progress and green transition, China and ASEAN in have great room for cooperation in air pollution prevention and control, water pollution prevention and control, soil pollution prevention and control and solid waste disposal. This section describes the efforts of China and ASEAN to address a series of emerging environmental problems, such as prevention and control of air pollution and water pollution.

### 5.2.9 *Air Pollution Prevention and Control in the Beijing-Tianjin-Hebei Region of China*

Regional air pollution characterized by $PM_{2.5}$ has become the most urgent and prominent environmental problem in the haze-prone Beijing-Tianjin-Hebei region. Motor vehicle emissions are an important source of air pollutants, while thermal power, iron and steel, nonferrous metals, petrochemical, cement, and chemical industries are the main industrial sources. On June 14, 2013, the State Council issued the Action Plan on the Prevention and Control of Air Pollution, which include ten measures for the prevention and control of air pollution to strictly control high energy consumption, limit newly added energy consumption in high polluting industries, quicken the pace of energy structure adjustment, and reduce pollutant emission. At the same time, China established regional joint control mechanism, strengthened performance assessment system for air environment governance in different provinces and cities as well as population intense regions and established credit system for the prevention and control of air pollution.

**Box 5.10 China's Action Plan on the Prevention and Control of Air Pollution**

Specific targets: By 2017, in the cities at prefecture level and above, the inhalable particulate matter concentration shall be decreased by 10% or more against 2012 level and the number of days with excellent or good weather conditions shall increase gradually; The fine particulate matter concentration of Beijing-Tianjin-Hebei region, Yangtze River Delta, and Pearl River Delta shall be decreased by 25, 20, and 15% respectively, among which the average annual concentration of fine particulate matter in Beijing shall be controlled at around 60 $\mu g/m^3$. To realize these targets, the Action Plan determined ten specific measures:

- Strengthen the comprehensive management to reduce the emission of multi-pollutants; renovate coal-fired small boilers and speed up desulfurization, denitrification, and dust removal renovation projects in key industries; control dusts in cities and the pollution from catering industry; speed up the phasing out of yellow label vehicles and out-of-date vehicles; develop public transit system, promote new energy vehicles and improve fuel quality.
- Adjust and optimize industrial structure, and promote the transformation and upgrading of the economy; strictly control the newly added production capacity in high energy consuming and high polluting industries; speed up the phasing out of backward capacity and stop the construction of illegal projects in seriously over capacity industry.
- Speed up the technological upgrading of enterprises, and improve their innovation capacity; develop circular economy and foster and develop environmental protection industry; promote the R&D and industrialized application of key environmentally friendly equipment and products.
- Adjust the energy structure and increase the supply of clean energy. By 2017, coal shall take less than 65% in the overall energy consumption mix. Beijing-Tianjin-Hebei, Pearl River Delta and Yangtze River Delta regions shall witness negative growth of their coal consumption.
- Increase the environmental requirements for market access of investment projects, optimize industrial spatial layout and strictly limit the energy-consuming and highly-polluting projects in ecologically vulnerable and environment sensitive areas.
- Give full play to market mechanism and improve environment economic policies. The central finance will set up special fund and substitute subsidy with reward. Adjust and improve the policies on pricing and taxation and encourage private and social fund to get into the air pollution prevention and control areas.
- Improve the legal system and tighten the supervision and management according to law. The central government will release air quality ranking in key cities in China regularly and set up mandatory information

disclosure system for major polluting enterprise. Improve environment supervision capacity and strengthen environment enforcement.

- Establish regional coordination mechanism and harmonize regional environment governance. Beijing-Tianjin-Hebei, and Yangtze River Delta need to establish coordinated working mechanism for air pollution prevention and control. The State Council will sign objective-based letter of responsibility with provincial government and will conduct annual assessment and practice strict accountability.
- Establish monitoring, early warming and emergency response system, develop and kick off the emergency response plan timely to appropriately address heavily polluted weathers.
- Identify the responsibility of different parties and mobilize wide public participation to improve air quality together.

According to the 12 Measures to Implement the Spirit of the Meeting on the Mechanism for Joint Prevention and Control of Air Pollution in the Beijing-Tianjin-Hebei Region and the Surrounding Area, pollution control actions shall follow the law and be science-based, and the focus will be put on non-point source pollution in such key industries as steel and thermal power. An online monitoring network will be built for enterprises with 90% or more industrial coal consumption in the region. The national strategy for coordinated development of Beijing, Tianjin and Hebei draws unprecedented attention to environmental protection. The three places have successively announced air pollution source apportionment results, and taken measures to promote local and regional environmental action plans, including innovative pollution control initiatives and strengthened law enforcement. Owning to the concerted efforts, the air quality has been improved remarkably. In 2014, Beijing-Tianjin-Hebei area achieved remarkable progress in air pollution prevention and control. 13 cities have an average of 156 days with the air quality up to the standard, 85 days less than the average of 74 cities, but 5.3% points higher than 2013. In 2015, the average concentration of PM2.5 in Beijing decreased by 6.2% on a year on year basis, hitting the pre-set target. In Tianjin, in the future 5 years, it will further reduce the consumption of coal and is expected that the concentration of PM2.5 will be reduced by 25% by 2020. The concentration of the six major pollutants in Hebei affecting the air quality have all be reduced dramatically, among which PM2.5 reduced by 18.9% than 2014 and 28.7% than 2013, hitting the target of 25% reduction by 2017 two years ahead of schedule.

**Box 5.11 Measures for air pollution prevention and control in the Beijing-Tianjin-Hebei region of China**

Beijing

- Put forward the clean energy strategy to build an energy system featuring substitution of clean energy for coal; and
- Introduce the Regulations of Beijing for Air Pollution Prevention and Control to strengthen local environmental protection.

Tianjin

- Perform strict law enforcement according to the principles of defined area, personnel, responsibilities, standards, process, and assessment; and
- Punish and shut down business involved in environmental violations, and cooperate with the public security departments in handling environmental criminal cases.

Hebei Province

- Optimize the industrial structure by reducing iron, steel, cement and plate glass production capacity, and organize crucial actions to control air pollution in steel, cement, electricity, and glass industries; and
- Implement desulfurization, denitrification and dust removal renovation projects.

The newly amended Air Pollution Prevention and Control Law, released on August 29, 2015, makes more stringent requirements on heavy fines for excessive emissions, clean coal utilization and key monitoring scope. Beijing, Tianjin and Hebei intensify air pollution control by promoting industrial upgrading, controlling energy-consuming industries and improving fuel quality. The solution to regional air pollution requires regional joint prevention and control based on close integration of scientific research and policy regulation.

### 5.2.10 Water Pollution Prevention and Control in China

China is in shortage of water resources and listed by the United Nations into the 13 water-poor countries. Currently, more than 400 of China's 669 cities suffer water shortage, 110 cities have severe water shortage, and more than half of sections in the country's seven major river systems have been polluted. Water pollution has risen as one of the major environmental problems. The protection of water resources and control of water pollution is considered as a necessary condition for sustainable development of medium-sized and large cities.

Liaohe river basin is one of the seven major river basins in China, crossing Hebei, Jilin, Liaoning provinces and Inner Mongolia (autonomous region) and was listed as one of the key rivers to be harnessed by the central government in 1996. Under the principle of controlling the source, intercepting pollution, restoring the ecology and enjoying prioritized development and with strong support from central finance, the pollution control of Liaohe river basin conducted a total of 438 projects with 87.6 billion yuan of investment into the major projects. Liaoning Province also drafted and issued 12 local laws and regulations, including Regulations for the Prevention and Control of the Water Pollution of Liaohe River Basin and Regulations for Liaohe Protection Zone in Liaoning. It released Sewage Discharge Standard in Liaoning Province which is even stricter than national standard, closed 417 paper mills and established large-scale integrated paper making industrial parks with high standard. Shenyang city closed all the sewage drainage exits in the city section of Huihe River totaling 32 km, cleared up 123 sandpits and demolished 58 illegal small oil refineries. During the 12th Five-Year-Plan period, the trunk stream of Liaohe River for the first time saw no water body with the water quality inferior to Class V and become the first batch of demonstration area for ecological civilization. Monitoring data show that the number of the species in Liaohe River Reserve has increased dramatically with better water quality and eco environment.

The Government of Dali Prefecture signed the letter of responsibility for Erhai Lake protection with principal leaders of eight departments of the Dali Prefecture, Dali City, Eryuan County. The targets and tasks are decomposed and risk capital mortgage and one ballot veto implemented to ensure positive results of water pollution control in Erhai Lake. It is also very critical to raise public awareness of environmental protection in addressing environmental problems of urban lakes. A key factor to success in governing the Erhai Lake lies in knowledge popularization, awareness enhancement, and good habits of the local residents.[18]

Weinan municipal department for environmental protection developed the Implementation Plan for the Three-Year Action to Control Water Pollution in the Weihe River Basin in early 2012, clarifying deadlines and tasks for water pollution prevention and control. The plan set down 60 governance projects involving an annual investment of 710 million yuan. The Environmental Protection Century Campaign centering on key governance projects (see Table 5.5) injected a strong impetus to water pollution control in the Weihe River Basin. In addition, the special rectification of sewage outfall was carried out and the water pollution compensation system introduced. In February 2013, the COD concentration at the exit section of the Tongguan Bridge on the Weihe River was reduced to 18 mg/l and ammonia concentration to 2.073 mg/l, down by 18.2 and 41.4% respectively.[19]

The governance of water pollution is of high importance and necessity, facing the shortage of water resources. In this respect, the effective policies and measures include: (a) strengthening supervision by relevant regulatory authorities;

[18]Source: China Environment News.

[19]Source: Weinan government portal, http://www.weinan.gov.cn/.

**Table 5.5** Environmental protection century campaign in the Weihe River basin

| Activity related projects | State | Number |
|---|---|---|
| Garbage disposal projects | Newly built | 1 |
| Rural pollution control projects | Completed | 6 |
| Nitrogenous fertilizer companies | Closed | 2 |
| Waste water treatment projects | Transformation for dephophorization and denitrification | 12 |
| 31 industrial waste water deep treatment projects | Shut down | 1 |
| | Put into operation | 12 |
| | Commissioning | 13 |
| | Under construction | 5 |

(b) enhancing people's awareness of water pollution prevention and control; (c) defining the responsibilities of the government and relevant departments; (d) introducing a harmonized management mechanism; and (e) making necessary amendments to laws and regulations.

### *5.2.11 Soil Pollution Prevention and Control in China*

As a result of fast industrialization, soil pollution is very serious in China now with 20 million hectares of arable land suffering heavy metal pollution, accounting for about 1/5 of the total, among which 2 million hectares suffer mineral pollution, 5 million hectares are polluted by petroleum and 50,000 ha by solid wastes. Industrial wastes polluted around 10 million hectares of land, and sewage irrigated area reached 3.3 million hectares.

In around 2010, soil remediation began to be promoted under the push of the central government. The Action Plan for the Prevention and Control of Soil Pollution laid out the action plans for soil environment protection and comprehensive control during the 13th Five-Year-Plan period. Most private environment protection enterprises have also joined the efforts to restore soil. Yongqing Environment Protection Group is the first listed company with successful cases for soil remediation. It independently developed the remediation technology for cadmium polluted soil with 20 million yuan of investment in two years. The cadmium content in rice was reduced by 60% by December, 2015. High Energy Environment is a company with "isolation barrier + environment remediation" as the core technology. With no appropriate commercial models for soil remediation in China now, this technology can better balance the dilemma of environment pollution treatment and funding shortage. Bossco Environment Protecting Technology Co., Ltd conducted a series of research and development efforts on soil remediation and heavy metal pollution and possessed the technologies of integrated thermal absorption, solidification, soil-free reclamation, and soil washing with rich experiences in project implementation. It established benchmark effect in the soil remediation project for the relocated plot of Nanning Chemical Group.

In order to control heavy metal pollution from the sources and reduce the emission of heavy metal, Hunan Province promoted clean smelting techniques including normal pressure and oxygen-rich leaching process for zinc residue and heavy metal pollution control technologies for zinc smelting industry. It also phased out and closed 1018 enterprises. In 2013, its mercury, cadmium, lead, chromium, and arsenic concentrations in industrial waste water were reduced by 11.94% from 2007 levels. The trunk stream of Xiangjiang River saw a decline of the concentrations of five heavy metals by 22.2–58.3%. Technologies favored for heavy metal pollution control also include quality-based reuse and integration of nonferrous metal smelting wastewater, in-depth electrochemical treatment, and heavy metal waste recycling. In addition, the colleges and universities, research institutes and enterprises in Hunan Province have reaped fruitful results in the research and development of soil remediation technologies applicable to heavy metal contaminated sites.[20]

Soil heavy metal pollution control projects are costly and the desired economic benefits are hardly achievable, so government regulation is considered as an effective measure. Heavy metal pollution control depends on the government's efforts in many ways: (a) developing targeted standards that reflect the characteristics of pollution, such as hidden risk, long duration, and wide range and at the same time; (b) taking a set of effective measures covering government management, policy implementation and government regulation. In addition, China's soil remediation market is open with different international cooperation and exchanges. It is expected to draw on good practices, technologies and equipment of foreign companies.

### *5.2.12 ASEAN Cooperation on Transboundary Haze Pollution Control*

In recent decades, transboundary pollution has stood out as a most important environmental issue in ASEAN. Haze pollution resulting from land and forest fires has exerted a serious negative impact on the ASEAN Member States, especially Brunei, Indonesia, Malaysia and Singapore. The health of local residents is threatened, as the regional air pollution index reached dangerous levels higher than 500.

To this end, since December 1997, the ASEAN Member States have worked together to monitor, prevent and mitigate the haze pollution under the guidance of the Regional Haze Action Plan (RHAP). In June 2002, the ASEAN Agreement on Transboundary Haze Pollution (AATHP) was signed by the Southeast Asian

[20]Source: China Environment News, http://www.cenews.com.cn.

countries. In addition, the ASEAN Transboundary Haze Pollution Control Fund was created with contributions of the ASEAN Member States on a voluntary basis. As a legal embodiment of the guiding spirit and specific measures of RHAP, the AATHP has achieved initial progress. As shown in Fig. 5.3, fire incidents in Indonesia between 1997–2011 decreased by 85%, but due to unbalanced responsibility and benefits concerning cross-boundary haze pollution prevention and control as well as certain domestic factors, not until it ratified RHAP in September 2014, the transboundary haze issue began to see more substantial progress.[21] At the same time, ASEAN Peatland Management Strategy (APMS) started to be implemented and "zero burning" and controlled burning was massively promoted. Meanwhile, Indonesia, Malaysia, the Philippines and Viet Nam are also actively working out a national action plan on sustainable peatland management to minimize peatland fire risk. They established fire monitoring and reporting system and rated the fire disasters in the region with rating assessment to minimize risks and reduce haze as a result of the fire [5].

**Box 5.12 Efforts made by Indonesian government to control haze and their difficulties**

Major measures:

- Prohibit burning with clear legal provisions;
- Issue relevant rules and regulations and organize local governments to study and discuss;
- Recall the permits from some forestry companies, arrest and punish the burners;
- Agriculture departments provide seeds, fertilizer and technologies to encourage people and companies to use other methods to clear forests for farmland; and
- Require relevant departments to distinguish fire if there are any.

Difficulties:

- Poor law enforcement from local governments;
- "Slash-and-burn" is a local way of life and is hard to be completely stopped in a short period of time;
- Due to financial and equipment limitations, the local farmers cannot afford the cost of opening up land with machines; and
- Short of sufficient forest protection and fire control personnel with insufficient firefighting equipment and transportation tools.

AATHP is a legal-binding obligation for individual countries to take actions. Indonesia also seeks regional cooperation to solve the issue of trans-boundary haze.

[21]Source: http://haze.asean.org.

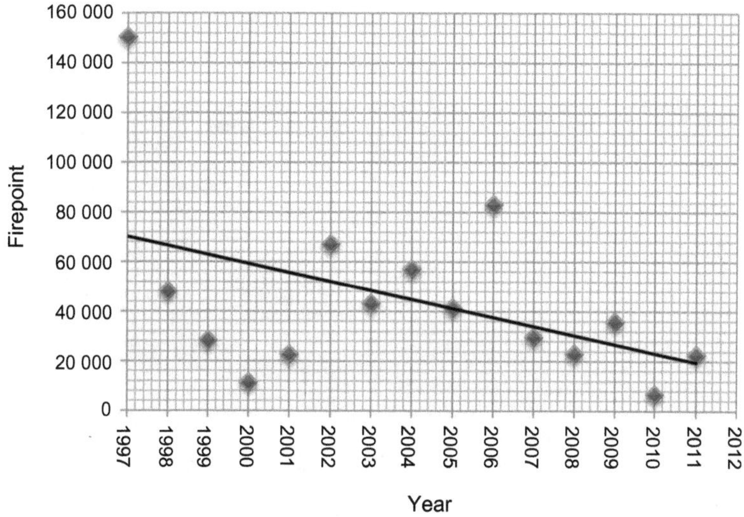


**Fig. 5.3** Trend analysis of fire incidents in Indonesia

Other ASEAN Member States have provided necessary assistance to Indonesia to improve its capacity to address the issue. The implementation of RHAP provides good reference for transboundary pollution prevention and control.

### *5.2.13 Singapore National Bicycling Plan*

For a 700 $km^2$ city-state like Singapore, "few roads relative to population" is a major problem challenging public transportation management. In order to ensure smooth flow of traffic and reduce air pollution caused by exhaust emissions, the Government of Singapore has been promoting green travel using a multi-pronged approach, such as control of private cars by ways of time-based road tolls and monthly public auction of car ownership certificate according to market supply and demand. To make riding safer, besides existing routes, new cycling paths, including the Rail Corridor and Round Island Route, are built to form an extensive network of off-road cycling paths which will facilitate the move of cyclists and green the whole community. The National Bicycling Plan aims to build a cycling network on the existing and planned cycling paths and park connectors which will allow residents to cycle safely from their homes to major transport hubs and key amenities. By 2015, 100 km of intra-town cycling paths will have been developed and another 90 km of cycling paths will be added to more remote towns by 2020.[22]

[22]Source: http://www.ura.gov.sg/MS/DMP2013/key-focuses/transport/cycling-for-all.aspx.

With emphasis on the social benefits of environmental protection system, Singapore continues to advance this beneficial systemic program adapted to local conditions through science-based planning. It provides better facilities and more attentive services to guide green travel, setting an excellent model for improving the environment and promoting sustainable urban development.

### *5.2.14 Decentralized Wastewater Treatment Systems (DEWATS) in Remote Areas of Cambodia*

Water is one of the most serious health problems facing the remote areas of Cambodia. Given the old habit of open defecation, the direct entry of human waste into the environment has led to water and land pollution and caused various diseases. The United Nations estimates that in 2008, only 23% of rural residents and 82% of urban residents can have access to improved sanitation. Bremen Overseas Research and Development Association (BORDA) launched the decentralized sanitation project to improve the living conditions of vulnerable groups and protect the environment. In cooperation with BORDA, Cambodia introduced the DEWATS system and first applied it to Sovann Komar. DEWATS consists of primary depositor, anaerobic baffle reactor, anaerobic filter, horizontal gravel filter, and discharge pipe. The system installed in the orphanage can process 15 $m^3$ of sewage per day and has such advantages as resistance to unstable flow, low maintenance costs, and high durability. The treated water meets the discharge standards for COD and BOD. Through this project, the sanitary conditions for about 120 orphans and orphanage staff have been improved and the pollution of the Mekong River alleviated accordingly. This contributes to the reduction of health risks of downstream residents, environmental threats and river body security risks.[23] The project involves low investment and maintenance costs and makes full use of local resources for effective operation. It can be extended in a large number of developing and developed countries.

### 5.2.15 *Semakau Landfill in Singapore*

Located 8 km south of Singapore, Semakau Landfill is the world's first offshore landfill involving an investment of 610 million Singapore dollars. A 7 km perimeter rock bund was built to enclose an area of 3.5 $km^2$. With a design life of 30 years, the landfill can accommodate 1500 tons incinerable waste and 500 tons non-incinerable waste every day, and therefore it can meet the waste disposal needs of Singapore until 2040 according to current trend. In addition, 136,000 $m^2$ mangroves are planted surrounding the landfill. The agreeable construction program guarantees scientific, hygienic and odorless landfill and ensure protection of the surrounding ecological environment.

To make the public aware of the importance of waste disposal and better protect the environment, the National Environment Agency of Singapore has taken a series of measures in Semakau Landfill to improve environmental publicity. The daily amount of waste disposal is reduced to 5000 tons from 7000 tons in the 1990s. Equipped with new solar and wind power systems, the beautiful environment of the island also attracts more and more people for birdwatching and mangrove sightseeing.[24]

This offshore landfill project gives inspirations to countries by showing that the landfill can be built into a recreational scenic area with the function of waste disposal. The efforts to enhance environmental awareness are also conducive to fundamentally reducing waste. Ecologically sustainable development needs more such projects.

## 5.3 Biodiversity Conservation

Biodiversity conservation is a major issue facing mankind and requires the joint efforts of countries around. Fully recognizing the importance of cooperation, China and ASEAN have actively advanced cooperation in the field of biodiversity. The two sides signed a number of strategic plans during 2009–2011 (see Table 5.6).

[23]Source: http://www.iwawaterwiki.org.

[24]Source: Ministry of the Environmental and Water Resources, Singapore.

**Table 5.6** China-ASEAN strategic plans in the field of biodiversity

| Time | Event |
|---|---|
| October 2009 | China-ASEAN strategy on environmental protection cooperation |
| October 2010 | Joint statement on sustainable development |
| September 2011 | China-ASEAN environmental cooperation action plan |

The ASEAN Center for Biodiversity (ACB) has held multiple biodiversity-related training programs and seminars, and awards the Certificate of ASEAN Heritage Park to qualified national parks which totals five in Viet Nam. The international community has highly appreciated these efforts and the EU and dialogue partners hope to establish cooperative relations with the center. This section describes the efforts of China and ASEAN in the conservation of biological diversity.

## *5.3.1 Nature Reserves in Sichuan Province, China*

Wolong National Nature Reserve, with a total area of 200 km$^2$, is located in the west edge of Sichuan Basin. There are diverse ecosystems covering almost all vegetation types from subtropical to periglacial environments and rich species which number at least tens of thousands with increase every year. Without undermining biological diversity, environment and recycling, Wolong National Nature Reserve takes an active approach of protection, through planned and rational use of resources. This increases income and reduces the financial burden, which properly solves the funding shortage. At the same time, the local government provides education to residents and assists the nature reserve in investigating illegal activities in the nature reserve [6].

**Box 5.13 Biodiversity conservation measures taken by Wolong National Nature Reserve**

- Continue to carry out biodiversity surveys;
- Improve protection mechanism and enhance protection inputs;
- Strengthen publicity and education, improve rules and regulations, and raise public awareness of protection and the legal sense; and
- Improve livelihood to enhance public willingness for and participation in biodiversity conservation activities.

Laojunshan National Nature Reserve, located in Pingshan County, Yibin City, Sichuan Province, covers a total area of 3500 ha. It is established mainly to protect such pheasants as Sichuan Hill Partridge (Arborophila rufipectus, endemic to China and listed as globally endangered by IUCN), Lady Amherst Pheasant

(Chrysolophus amherstiae), Silver Pheasant (Lophura nycthemera), Tragopan Temminckii and the associated rare wildlife and sub-tropical broadleaf forest ecosystems. Given rich biodiversity and high conservation value of the nature reserve, the World Pheasant Association (WPA), German Zoological Society for the Conservation of Species and Population, and North of England Zoological Society (NEZS) assist raising funds and provide contributions to the infrastructure and personnel training. Laojunshan National Nature Reserve Administration provides management, and the Pingshan County Forestry Bureau takes charge of construction. Laojunshan National Nature Reserve is also establishing itself into a wild giant panda habitat, owning widely distributed bamboos favored by giant pandas, such as Arrow Bamboo (Pseudosasa japonica), Chimonobambusa szechuanensis, and Yushania mabianensis. These diverse bamboos extend to perimeter protection forests to more than 100,000 μ, and provide ample food for wild giant pandas. In 2015, at the National Working Meeting on Giant Panda Conservation, the Laojunshan National Nature Reserve was officially upgraded to Giant Panda Habitat which with 3201 ha, is home to three pandas.[25]

Nature reserves established in China have played a very important role in in situ biodiversity conservation. They cover almost all important ecosystems in the country. With the expansion of protected areas and species habitats, many large animals will be able to move around in a larger range, thereby facing smaller pressure to survive. More species will be better protected with further consolidation and merger of protected areas.

### *5.3.2 Biodiversity Conservation Cooperation in ASEAN Region*

ASEAN is strengthening its protection of biodiversity and its member states are embracing deeper regional cooperation to conserve their rich biodiversity so as to consolidate ecosystem services in regional development of society, economy and environment.

**Box 5.14 Moves for biodiversity conservation in ASEAN region**

Regional guidelines and action plans:

- Regional action plan for ASEAN Cultural Relics Park in support of Programme of Work on Protected Areas (PoWPA);
- Regional action plan for Global Taxonomy Initiative (GTI) in support of GTI; and

[25]Source: http://www.scljs.org/.

- "Management guideline for transboundary protected areas" in support of PoWPA.

Thematic and cross-sectoral regional plans:

- Support capacity building on taxonomy and enhance the biodiversity knowledge of ASEAN region by training experts on taxonomy;
- Enhance the capacity building for access and benefit sharing (ABS);
- Improve public awareness of endangered species, reinforce law enforcement, better and maintain regional learning framework and the partnership with ASEAN-WEN and Wild for Life; and
- Other cross-sectoral moves, including strengthening biodiversity conservation and climate change mitigation, promoting the participation of the commercial sector, filling the information gap on biodiversity, enabling more stakeholders to understand the value of biodiversity via different media outlets and shaping strategic partnership.

Information exchange mechanism:

- Compile auto abstract of different species;
- Establish invasive alien species (IAS) database with input channels;
- Establish E-library which can be downloaded to tablet computer or mobile phone;
- Develop apps for "ASEAN Cultural Relics Park"; and
- Interactive mapping tool to map the appearance of different species in the protected areas according to the data of IUCN and World Database on Protected Areas (WDPA).

The ten ASEAN Member States are all the contracting parties of Convention on Biological Diversity (CBD). They are committed to strengthening biodiversity conservation measures and raising public awareness of biodiversity's values via media outlets and education. They cooperated with other countries and international organizations to develop education and awareness raising plans on biodiversity conservation and sustainable use. In 2005, ASEAN Member States established the ASEAN Center for Biodiversity and worked toward taxonomy, information management, biodiversity assessment and management. Their efforts increased the knowledge of the public on the value of biodiversity, and reduced its loss rate and thus won international recognition. In addition, ASEAN established ASEAN Working Group on Nature Conservation and Biodiversity (AWGNCB) which served as a consulting platform for specific actions and committed to establishing a sharing platform in ASEAN region. Under the cooperative framework of regional biodiversity conservation, ASEAN established a great number of national natural resource protected areas, national parks and genetic conservation areas. Brunei,

Indonesia, Malaysia, the Philippines, Singapore, Thailand and Viet Nam established national departments/units for biodiversity conservation.

**Box 5.15 Work content of ASEAN Center for Biodiversity (ACB)**

- Develop and disseminate materials and knowledge products for communication, education and public awareness;
- Share good cases on communication, education and public awareness;
- Use extensively internet and social media for visibility;
- Report ACB projects sponsored by donors;
- Report important ACB activities;
- Hold publicity activities:
- Hold activities of "United Nations Decade on Biodiversity", "International Day for Biodiversity" and "Global Green Wave"; and
- Establish "ASEAN Biodiversity Leading Award" to recognize the youth teams, media and commercial sectors who made individual or collective contribution to the protection and sustainable development of biodiversity in ASEAN region.
- Forge partnership with government, private sectors, media and schools;
- Make use of the research results of government and non-governmental organizations; and
- Cooperate with ASEAN-WEN to prevent illegal hunting of wild animals in Southeast Asia.

The cooperative mechanism for biodiversity conservation in ASEAN region serves as a scientific model for developing and implementing biodiversity cooperative projects and is also an excellent example for biodiversity conservation and regional ecological safety among different regions.

### 5.3.3 *Coral Triangle*

The Coral Triangle is located in the tropical marine waters of six countries, namely Indonesia, Malaysia, Papua New Guinea, the Philippines, the Solomon Islands and East Timor. The triangle area, approximately 600,000 km$^2$, accommodates more than 500 species of reef corals and 3000 species of fish.

The Coral Triangle is currently facing such threats as unsustainable fishing, waste pollution, coral mining, and global warming. In order to ensure sustainable development, the heads of state of above-mentioned six countries signed the Coral Triangle Initiative on Coral Reefs, Fisheries and Food Security on May 15, 2009, strongly demonstrating the national commitments to conservation of the Coral Triangle and sustainable management of marine, coastal and island ecosystems. The move aims to secure a better future for the people of the Coral Triangle region. The triangle area also bears obligations for the greatest health of global marine organisms and the protection of plants, animals and ecosystems.[26]

The Coral Triangle provides a good model of regional cooperation to protect the world's scarce resources. It embodies the unremitting efforts of countries to preserve resources and heritage and secure the demand and rights of future generations. The environmental issue is like the center of the circle which cannot be isolated from the development of all countries and can be well addressed only through joint effort.

### 5.3.4 Heart of Borneo (HoB)

Located in the heart of the world's third largest island Kalimantan (also known as Borneo), the Heart of Borneo is one of the world's most important diversity centers. Covering an area of 234,000 $km^2$, it accommodates about 6% of the world's biodiversity resources, including 13 species of primates, 150 species of reptiles and amphibians, 350 species of birds and 15,000 species of plants. Yet there are always new species to be discovered. In addition, the Heart of Borneo is one of the world's only two woodlands with chimp, elephant and rhino communities.

However, more than half of the primary forests in Borneo have been lost over the past three decades due to illegal logging, illegal hunting, deforestation for oil palm plantation, and land conversion for plantation and mining. In order to achieve sustainable development, the World Wide Fund for Nature (WWF) launched the Heart of Borneo Initiative. The Heart of Borneo Declaration was signed by Indonesia, Brunei, and Malaysia on February 12, 2007, in order to ensure the effective protection of forest resources and diversity of species in the region. The three countries committed to the establishment of a transboundary network of protected areas, including nature reserves, production forests and land for other purposes. The initiative encompasses five components: transboundary management, protected area management, sustainable management of natural resources, ecological tourism development, and capacity building.[27] According to the Environmental Status Report of the Heart of Borneo 2014 issued by WWF, the Heart of Borneo Declaration has produced desired results. During 2007–2012, the deforestation rate was 4.68% in Borneo and only 2.19% in the Heart of Borneo. The water quality of three water areas in six water basins in the Heart of Borneo was satisfactory, very good in one and good in other two. Currently, 88% of the historical plant species are remained in the Heart of Borneo.

Initiated by NGOs, this cross-border biodiversity conservation initiative, in collaboration with governments, is of positive significance for strengthening inter-governmental cooperation, protecting regional ecological environment,

[26]Source: http://www.jsof.gov.cn/art/2009/6/29/art_127_38808.html.

[27]Source: http://wwf.panda.org/.

promoting sustainable development, and addressing global climate change. It is expected to become the new hot spot of region cooperation.

### 5.3.5 Green Contract for Communities Living in Buffer Zone of U Minh Thuong National Park in Viet Nam

U Minh Thuong National Park (UMTNP) is one of Viet Nam's two largest peat swamp forests and recognized as one of the three priority areas for wetland conservation in the Mekong Delta. UMTNP covers an area of 21,800 ha and consists of a core zone and a buffer zone. The core zone, as large as 8509 ha, contains rich flora and fauna species, including 250 species of plants, 32 species of mammals, 186 species of birds, 39 species of reptiles and amphibians and 34 species of fish, of which 40 species are included in Viet Nam Red Book or IUCN Redbook.

During the dry season in March 2002, totally 2700 ha of melaleuca forests in UMTNP were lost due to fires. To manage UMTNP sustainably, the Government of Viet Nam signed green contract with communities living in buffer zone in 2012, according to which, residents enjoy the benefits of livelihood development projects and the government provides agricultural incentives, such as rice, vegetables and pitaya. The members to the agreement shall (a) not participate in illegal activities of hunting, fishing, logging, honey harvesting and burning inside the UMTNP; (b) have the responsibility to dissuade and prevent others who intend to enter the park illegally; (c) participate in the activities of forest fire management of national parks and their buffer zones; (d) not engage in trading activities of wild animals prescribed by the wildlife management agencies, particularly species listed in IUCN Redbook and VietNam Red Book.[28]

UMTNP presents good social and ecological effects by combining ecological and social functions. Establishing national parks is an important measure to protect the ecological environment, presenting a combination of ecological and social functions. The management measures taken by the authorities also mobilize the local residents.

### 5.3.6 Nature Society of Singapore

The Nature Society of Singapore (NSS) is a non-governmental and non-profit organization dedicated to the appreciation, conservation, study and enjoyment of the natural heritage in Singapore, Malaysia and the surrounding region. NSS (a) organizes nature appreciation activities like guided nature walks, bird and butterfly watching, talks and overseas eco-trips; (b) conducts conservation projects

[28]Source: http://kiengiangbiospherereserve.com.vn/.

and surveys; (c) collaborates with schools and community groups to promote nature appreciation and education; and (d) campaigns for the protection of natural habitats. The achievements of NSS include (a) successfully persuaded the government to set aside Sungei Buloh as a mangrove and bird sanctuary which is now legally protected and known as Sungei Buloh Wetlands Reserve; (b) relocated coral reefs which were threatened by land reclamation; (c) staved off plans to develop part of Peirce Reservoir forest into a golf course; (d) published the Master Plan for the Conservation of Nature in Singapore, which has influenced government policy and planning; and (e) advocated the preservation of Chek Jawa, a unique marine habitat on Pulau Ubin. The Nature Reserve Management and Development Plan for Singapore published by NSS also exerts a major impact on government decisions.[29]

NSS has been accorded the status of an Institution of Public Character, in recognition of its contribution to biodiversity conservation. Given NGOs' undeniable role in the protection of natural ecosystems and biodiversity, the governments should facilitate the full play of civil societies like NSS, in order to create a favorable social atmosphere. These organizations can offer comments or suggestions on government decisions and provide education services to the general public.

## 5.4 Features and Outlook

As developing or newly industrialized countries, China and ASEAN Member States are facing many common challenges in the field of environment and development. They have realized that the green economy will become the new engine of economic development, and developed nationally appropriate green development planning and policy implemented in the country mainly through collaboration between local governments and enterprises, taking into account the historical background, natural conditions and specific environmental problems encountered in reality. These measures and policies are distinct due to different stages of development of countries, but the common characteristics and laws embodied in green practices provide a reference for other countries and regions.

### *5.4.1 Multi-stakeholder Participation in Green Development Measures*

The cases selected in this chapter imply that the environmental governance system should involve all stakeholders, including government, private sector and the general public. The smooth implementation of a policy or a project requires not only the government's guidance and supervision, but also the active participation of

[29]Source: http://www.nss.org.sg/.

enterprises. It also needs public acceptance and support, assistance of NGOs, or even linkage between regions and between countries. Green development is the transformation and upgrading of traditional economic development mode. It requires government policy guidance and support which includes regulations, programs, laws specific to problems and local conditions, policy support and incentives granted for green development projects, and restrictions and punishments on projects contrary to green development. Policy is particularly important for green development. An effective policy based on national realities can encourage individuals and enterprises to change the production and consumption patterns, and maximize social and economic benefits while safeguarding sustainability of environmental resources. In addition, the efforts of every citizen are indispensable, and through awareness raising and education, the idea of green development should be conveyed to the public and integrated into domains that policy cannot reach. To achieve desired results in coordinated economic, social, and ecological development, the effective move for governments is to seek and support cooperation that leverages social and international resources and mobilizes multiple stakeholders for green development. Stakeholders involved in green development are both practitioners and beneficiaries. In the participatory process, stakeholders should perform respective duties according to coordinated rights and obligations, in order to ensure long-term sustainability of development.

### *5.4.2 Technological Innovation to Drive Green Development*

The importance of technological innovation for enterprises to achieve green development can be observed in such cases as China's automobile sector, SCG, and DEWATS. Enterprises are the major actors to directly translate green ideas into products and services that meet consumers' demand. They serve as the primary driving force to a green economic transition, including greening production and service process, launching green products, and developing environmental technologies. With continuous technological progress, innovation and high-tech driven environmental protection and green development has become a global trend. In this context, it is also very important to promote social and economic development from a technical perspective.

### *5.4.3 Incentive Mechanism to Promote Green Development Projects*

Incentives and certificates to projects are conducive to expansion of projects or regional influence, as shown in the cases of biodiversity conservation and nature reserves development. These incentives inject a powerful impetus to the implementation of similar activities in the future by wining wider attention and recognition for projects. Environmental and economic benefits are by no means conflicting in the process of green development. In the long run, with policy support and self-value development, green projects can also make considerable earnings and profitable projects are more sustainable. Regarding eco-agriculture and eco-village development, measures for environmental protection and green development have brought considerable economic benefits for local communities, and enhanced grassroots participants, noticeably in remote areas and governance-weak areas. The incentive mechanism is particularly effective in the early stage of green development, especially for enhancing public participation.

### *5.4.4 Measures Adapted to Local Conditions to Ensure Implementation*

Considering differences in natural resources, geographical features, and cultural customs, the patterns of green development can be adapted to local conditions, with reference to each other based on common grounds. Eco-villages in Viet Nam accurately grasp the production capacity of local communities and natural conditions, while those in Thailand make full use of local coconut waste for biomass production. The projects of Coral Triangle and the Heart of Borneo established distinct biodiversity conservation strategies based on an accurate understanding of local natural resources. The one-size-fits-all approach to green development is inefficient and unsustainable.

Green development is not just a strategic issue at the national level, but also the priority for international cooperation. It will help to bring about new growth points for the global economy, improve people's living environment, and promote long-term regional cooperation and coordinated development. Green development needs the push from national government, joint efforts of different regions, and international cooperation. China and ASEAN share common objectives for greener development and have already accumulated certain experiences. Their environmental cooperation has been further expanded with new cooperative models. In the process of promoting green transition of the economy, China and ASEAN enjoy great potential for further cooperation in sewage treatment, air pollution prevention and control, clean production technologies and new energy development. Different countries shall enhance experience exchanges and forge partnership in the region for a win-win situation. In terms of policy, it is recommended to carry out joint policy research based on the exchanges of national experience in policy making. On technologies, actively participating in the new round of industrial and scientific revolution is suggested, with environment friendly technologies and industry as the starting point. On resources, it is important to expand financing channels, seek deeper and more solid cooperation on human resources, business opportunities and development model and thus establish multi-level cooperative platform and network in wider fields to shape efficient cooperation and sharing mechanism. On culture, it will be useful to promote green development concept and humanity communication.

Different countries can learn from each other on green development cases and experience, identify common problems and individual concerns. Developed countries are expected to share their experiences with developing nations and help them materialize technological upgrade for economic transition. Developing countries will actively learn from developed nations to improve their own situations. Developing countries can also learn from each other in the transition period for environment protection, share their excellent cases and seek common progress. To further deepen the cooperation and realize regional green development, China and ASEAN Member States are advised to take the opportunity of Belt and Road Initiative, build the Maritime Silk Road together and forge regional cooperative community for environment protection. Existing China-ASEAN cooperative platform and framework shall be further used to deepen the cooperation between China and ASEAN, and the regional cooperation in Greater Mekong Sub-region, promote the establishment of China-ASEAN cooperation demonstration base for environment protection industry and technologies and China-ASEAN maritime environment pilot zone. The concerted efforts of both sides will sure inject sustained driving force for the comprehensive regional cooperation, bring about new economic growth points, facilitate synergyistic and realize harmonized development of environment and economy.

## References


1. Suksri, P.: Sustainable Agriculture in Thailand, pp. 7–8. School of Business and Commerce, Keio University (2008)
2. Jianting, W., Yingying, L.: Study on the green development path of housing industrialization in China. Sci. & Techno. Prog. Policy **19**, 12–15 (2010)
3. Yuqian, W.: Thai's sustainable development in tourism and its lesson. Around SE Asia **10**, 41–44 (2006)
4. Mingjiang, D., Xiangquan, L., Aihua, L., Yi, Z., Dangxian, W.: Regulation of supply and demand structure of the water resources and support economic and social leap-forward development of projection measures. Arid Land Geogr. **5**, 379–390 (2011)
5. Xiaoyong, Cheng: Review and analysis on Southeast Asian trans-boundary Haze Governance. SE Asian Stud. **3**, 23–29 (2015)
6. Sihui, L., Jike, L., Yingxiang, W.: The relationship between Biodiversity conservation and nature reserve in China. World Forest. Res. **15**, 47–53 (2002)

# Chapter 6
# An Outlook into the Future: Toward Green Transition

**Abstract** At present, green development has turned into a new approach and engine for global sustainable economic growth. It is deemed as an effective strategy to realize sustainable development while alleviate long-term poverty, and a crucial measure for restructuring world economic order and global environmental governance. The UN Sustainable Development Summit in 2015 adopted the 2030 Agenda for Sustainable Development, which serves as a new global development agenda succeeding the Millennium Development Goals (MDGs). Paris Agreement adopted in the same year has laid foundation for the world economy to shift toward low carbon emission and climate change adaptation. Facing the future, China and ASEAN will embrace further boom in social and economic development with the changing demography, lifestyle, and access to basic services in Asia-Pacific region, and at the same time tackle challenging resource constraints and environmental pressures brought by further regional integration. It is thus acknowledged that green development and associated cooperation will be a promising choice for both sides. Inspired by the long-term goals of building a "beautiful China" and a "green ASEAN," respectively, and boosting green development with concerted efforts, China and ASEAN Member States are expected to choose effective policy measures to translate current national policies into practical actions to address dozens of challenges in environment and development field, including unbalanced development, environmental pollution and ecological degradation, weak governance capacity, insufficient financial resources, as well as inadequate technical measures and innovations. China and ASEAN have already initiated various cooperative efforts in the related fields. Such responses and actions as policy dialogue, joint research, capacity building, industrial cooperation, and pilot project are proven effective to deepen China-ASEAN cooperation in nine proposed priority areas. China and ASEAN will work closely to push forward green, low-carbon, and sustainable development in the region.



China-ASEAN Environmental Cooperation Center, *China-ASEAN Environment Outlook 1 (CAEO-1)*, https://doi.org/10.1007/978-981-10-6211-7_6

## 6.1 Global Context: Sustainable Development Gets into Implementation Phase

Although the poverty reduction-oriented MDGs have made historical contributions to the development of entire human race, ever since beginning of the new millennium, the international community is still challenged by the severe situations in terms of poverty alleviation and hunger eradication. At the same time, various non-traditional security issues, combining such global challenges as unbalanced development, climate change, energy security, food security, disaster prevention, and control impose growing threats on human existence and development. In September 2015, heads of governments around the world adopted the "Transforming Our World: The 2030 Agenda for Sustainable Development" at the UN Sustainable Development Summit, which serves as a new global development agenda succeeding the Millennium Development Goals (MDGs).

The 2030 Agenda for Sustainable Development highlights 17 Sustainable Development Goals (SDGs, with 169 associate targets). Compared with MDGs, this new agenda gives more attention to environment and brings the global recognition of environmental issues to a new height. It emphasizes an integrated approach towards the implementation of all the goals, with comprehensive solutions to economic, environmental and social problems. It conveys a strong message that environmental sustainability has become a key criterion to gauge development results and implementing SDGs has turned into a shared vision of the international community.

**Box 6.1 17 Sustainable Development Goals in the 2030 Agenda for Sustainable Development[1]**

Goal 1 End poverty in all its forms everywhere

Goal 2 End hunger, achieve food security and improved nutrition, and promote sustainable agriculture

Goal 3 Ensure healthy lives and promote well-being for all at all ages

Goal 4 Ensure inclusive and equitable quality education and promote lifelong learning opportunities for all

Goal 5 Achieve gender equality and empower all women and girls

Goal 6 Ensure availability and sustainable management of water and sanitation for all

Goal 7 Ensure access to affordable, reliable, sustainable, and modern energy for all

Goal 8 Promote sustained, inclusive and sustainable economic growth, full and productive employment and decent work for all

Goal 9 Build resilient infrastructure, promote inclusive and sustainable industrialization, and foster innovation

Goal 10 Reduce inequality within and among countries

Goal 11 Make cities and human settlements inclusive, safe, resilient, and sustainable

Goal 12 Ensure sustainable consumption and production patterns

Goal 13 Take urgent action to combat climate change and its impacts*

Goal 14 Conserve and sustainably use the oceans, seas, and marine resources for sustainable development

Goal 15 Protect, restore, and promote sustainable use of terrestrial ecosystems, sustainably manage forests, combat desertification, and halt and reverse land degradation and halt biodiversity loss

Goal 16 Promote peaceful and inclusive societies for sustainable development, provide access to justice for all and build effective, accountable, and inclusive institutions at all levels

Goal 17 Strengthen the means of implementation and revitalize the Global Partnership for Sustainable Development

*Acknowledging that the United Nations Framework Convention on Climate Change is the primary international, intergovernmental forum for negotiating the global response to climate change.

Due to its scale, ambition and wide coverage, the new Agenda calls for the joint efforts and full participation of countries all around the world to ensure its implementation, which will bring together governments, private sector, civil society, and

[1]*Source* UN Document A/RES/70/1, available at. http://www.un.org/ga/search/view_doc.asp?symbol=A/RES/70/1&Lang=E.

all other actors and mobilize all available resources to hit relevant targets. According to research of International Resource Panel (IRP), 12 out of 17 SGDs need to utilize the natural resources in a sustainable way in order to boost human well-being and 10 can only be realized when there is a substantial improvement of the utilization efficiency of land, water, energy (fossil and biomass), materials, and other limited resources. Goal 12–15 require to lessen the negative impact of economic development on ecosystems and their services.[2] SDG's implementation puts forward higher requirements for resources and environment governance in particular to China and ASEAN Member States which are still in the development stage and depend on natural resources to boost economic growth.

Climate action is a key part of the process to facilitate sustainable development. In December 2015, over 200 contracting parties of UNFCCC adopted unanimously the Paris Agreement which sets up new institutional framework and arrangements for global actions to fight climate change after 2020. In order to limit the rise of global temperature within 1.5 °C, countries in the world need to take low-carbon and green economy as their long-term development goals, further reduce carbon emission and enable a climate resilient economy. It sets the cornerstone for the world economy to move toward low carbon emission and climate change adaptation.

The two milestone events of 2030 Agenda for Sustainable Development and Paris Agreement show that the international community has already reached consensus on sustainable development with well-shaped philosophy, theory and corresponding institutional arrangements and is moving toward implementation phase. With an emphasis on coordinated development of economy, resources, and environment, green development provides an important policy option for the world to implement SDGs and take climate actions. In this context, China and ASEAN Member States will embrace more opportunities for their cooperation on green development.

## 6.2 China and ASEAN: Future Trends in Environment and Development

As ASEAN's biggest neighbor, China is connected to the area by lands and seas. In the history, Southeast Asia was an important hub on the ancient Maritime Silk Road. Over the past years since China and ASEAN opened dialogue in 1991, the two sides have increasingly support each other in development. Since the launch of China-ASEAN Free Trade Area (CAFTA) in 2010, China and ASEAN have been directed to build a more integrated region in terms of trade, investment, production, and infrastructure. In 2014, the bilateral trade between China and ASEAN reached 480.1 billion US dollars, with an 8.23% increase compared to the 2013 level; while the accumulated investment volume amounted to 120 billion US dollars. At present,

[2]UNEP, Policy Coherence of the Sustainable Development Goals: A Natural Resource Perspective. 2015.

China ranks as ASEAN's biggest trade partner, while ASEAN is China's third-biggest trade partner, fourth-biggest export market, and second-biggest import market. In 2015, negotiations on upgrading the largest free trade area for developing world yielded fruitful results, and the targeted trade volume of 1 trillion US dollars is expected to be achieved by 2020.

In light of global economic growth, Asia and the Pacific is viewed as the region with the fastest growth rate in the world. China and ASEAN are likely to become one of the important growth engines in Asia and the world. ASEAN has launched negotiations on the Regional Comprehensive Economic Partnership (RCEP), making crucial contributions to the regional integration. The ASEAN Community was formally established at the end of 2015 comprising the ASEAN Political-Security Community, the ASEAN Economic Community, and the ASEAN Socio-Cultural Community, marking a significant stride in ASEAN integration. Southeast Asia will be shaped into a single market with more than 600 million people and 2 trillion US dollars'worth of economic aggregate. Initiatives such as the 21st Century Maritime Silk Road and the Asia Infrastructure Investment Bank (AIIB) proposed by China will also reinforce regional connectivity and facilitate building a community of common destiny between China and ASEAN.

## 6.2.1 Forecast of Society and Development

The world economy is still in a sluggish recovery with the international trade slowing down. Though it dealt a short-term blow on the economy of Asia-Pacific region, Asia and the Pacific will remain as one of the most dynamic regions for world economy in the future. It is predicted that the economic growth of Asia-Pacific region will make greater contributions to world GDP with an expected share of 53% by 2050.[3] In this process, policy measures like structural reform will play a major role as they are conducive to the rebalance of demand and supply, reducing vulnerability, increasing economic efficiency, improving environmental quality, and ensuring Asia's leading role in world economic growth. China's economic transition will play a key part,[4] and ASEAN Economic Community will also push regional economic development.

The UN predicted a total population of 5.15 billion in Asia-Pacific region by 2050, and China will reach the peak of 1.416 billion people in 2029. After 2031, China's population will gradually shrink and drop to 1.348 billion by 2050. At the same time, the population of ASEAN Member States will continue to rise and is expected to reach 790 million by 2050.[5] In the future, the fertility and mortality

[3]The Economist Intelligence Unit. Long-term macroeconomic forecasts: Key trends to 2050. 2015.

[4]IMF. Regional Economic Outlook: Asia and Pacific. 2016.

[5]ESCAP. Statistical Yearbook for Asia and the Pacific 2015. 2015. http://www.unescap.org/resources/population-syb-2015.

rates in Asia and the Pacific will both see a downward trend which will result in a fast aging population and a reduction of labor force. Meanwhile, urban population will continue to rise and is expected to account for 63% of the total by 2050. People's lifestyle and their access to environmental services will all change as a result.[6] These trends will present themselves in different degree in China and ASEAN Member States.

#### 6.2.1.1 Economic and Social Development Forecasts of China

The future five years is a critical period for China to achieve its mid- and long-term socioeconomic development goals, and an important time span for China to address the middle-income trap. Such major efforts as building ecological civilization, strengthening pollution prevention and control, reinforcing environmental enforcement, and deepening institutional reforms would make substantial progress during this period of time. Thus, the future five years would be extremely important for China to boost green development.

While stepping into the new normal state, China's economic growth will switch from high speed to moderately high speed with more obvious trend of transitioning into a stage of higher state, more optimal labor division, and more rationalized structure.[7] The economic structure is supposed to upgrade to the mid- and high-end of the value chain from the current low- and mid-end. The industrial structure starts to turn from secondary industry-dominated pattern to tertiary industry-dominated pattern, while the demand structure changes from investment-dominated pattern to consumption-dominated pattern and consumption upgrading is accelerated. Consumption focus will change from subsistence to development. The economic driving force will gradually shift from element and investment to innovation. New type of industrialization, informationization, urbanization, and agricultural modernization are developing fast with more newer and stronger drivers for growth. Economic risks gradually reveal themselves in an obvious manner. Resources and environment constraints become intensified. At the same time, Chinese economy will further release its reform dividend, and the potential on domestic demand will be realized with stimulus from new-type urbanization and urban–rural integrated development. Major development strategies such as building "Silk Road Economic Belt" and "21st Century Maritime Silk Road" (the "Belt and Road Initiative"), Jing-Jin-Ji (Beijing-Tianjin-Hebei) Economic Integration and Yangtze River Economic Belt will boost regional economic development in China. It is forecasted that, by 2030, China could maintain its growth rate at 7.0%, with a per capita income of around 26 000 US dollars (PPP in 2005 terms).[8] Using 10 US dollars to

[6]UNEP. Global Environment Outlook (GEO-6): Regional Assessment for Asia and the Pacific. 2016.

[7]Outline for the 13th Five-Year Plan for China's Economic and Social Development. 2016.

[8]ADB. ASEAN, PRC and India: the Great Transformation. 2014.

100 US dollars per capita per day as a standard, about 79% of Chinese population will belong to the middle-income group by 2030.

China's social development will enter into a new stage, manifesting the following trends and features: fast-growing aging population, shrinking working-age population, and disappearing population dividend. At the same time, improved education coverage and quality, in particular those of higher education, further raises the quality of labor force. Income gaps between coastal regions and inland regions and between urban population and rural population are expected to be narrowed, but the provision of basic public services is still insufficient and the task to eliminate poverty is arduous. There will be increased public demands for high quality life, which will raise the share of consumption in GDP. The social interests and demands would become diversified.

The rapid urbanization in China will be accompanied with following problems, such as large scale and high speed expansion, unbalanced development, conflicting urban–rural structure, high intensity of resource/energy use, severe pollution and deterioration of ecological environment. China will continue to eye accelerated urbanization during the 13th Five-Year Period (2016–2020), by the end of which the permanent urban residents will account for 60% of the total population. At the same time, urban development would have a bettered overall layout, with improved quality. At the same time, China will establish a new and all-round opening-up pattern with Belt and Road Initiative taking the lead, strengthen South–South cooperation among the developing countries, expand foreign aid, and improve the quality and quantity of overseas investment to shape a cooperative pattern featuring openness and inclusiveness, high integration and mutual benefit and a new scenario for economic, trade and cultural exchanges.

#### 6.2.1.2 ASEAN Economic and Social Development Forecasts

ASEAN has an economy of 2.5 trillion US dollars in aggregate, ranking seventh in the world. It has signed free trade agreements with partner countries such as China, Japan, Korea, Australia, and India. The ASEAN Community was formally established on December 31, 2015 as planned, which includes ASEAN Political-Security Community (APSC), ASEAN Economic Community (AEC), and ASEAN Socio-Cultural Community (ASCC), with particular emphasis on AEC. The establishment of AEC is expected to shape a common market covering a population of 600 million with a total GDP of 2 trillion US dollars (source: World Economic Forum). Under the AEC blueprint, the AEC is built toward the following goals: (a) a single market and production base, (b) a highly competitive economic region, (c) a region of equitable economic development, and (d) a region fully integrated into the global economy.[9]

---

[9]http://www.asean.org/wp-content/uploads/2012/05/56.-December-2015-Fact-Sheet-on-ASEAN-Economic-Community-AEC-1.pdf.

It is reckoned that the average GDP growth rate of ASEAN will reach 5.4% by 2030, with the projected per capita income of around 12,000 US dollars.[10] The implementation of AEC, coupled with a shift in global manufacturing center as a result of rising Chinese labor costs, could create an annual economic value between 280 million US dollars and 615 million US dollars for ASEAN by 2030, which would be equivalent to 5–12% of ASEAN GDP in 2030.[11] At the same time, it is projected that the full implementation of AEC will generate by 2025 an additional 14 million jobs in six ASEAN economies (Cambodia, Indonesia, the Lao PDR, the Philippines, Thailand, and Vietnam), speed up the pace of structural change, and drive up productivity.[12] As a result of increased per capita income, the living standards for ASEAN population would rise substantially, with enhanced middle-income groups. According to a study from Asian Development Bank (ADB), against the standard of 10–100 US dollars per capita per day, around 64% of ASEAN population will belong to the middle-income group by 2030. Urban population is supposed to account for 56% of the total population in ASEAN by 2030.[13] Advantages in population trend, sustained urbanization, soaring productivity, and improved technologies will become major forces to fuel ASEAN's development.

### *6.2.2 Future Environment Outlook*

According to the forecasts made by UNEP's Global Environment Outlook 6 (GEO-6), with the changing demography, lifestyle and access to basic services in Asia-Pacific region, economic and social development will impose greater pressure on resources and environment which are specifically reflected in the following areas. First, increased resource consumption with low resource use efficiency will trigger extensive environment degradation, loss of ecosystem services, excessive wastes, and extra economic burden as a result. Second, natural disasters and extreme weather events caused by climate change will bring about bigger impact and economic losses. They may even offset the achievements of development and threaten the security of drinking water and food. Third, there are greater health risks due to the pollution of air, water, earth, hazardous chemicals, and heavy metals. In addition, researches also show that the emission of ozone and fine particulate matter (short-lived climate pollutants such as black carbon) is on a rising trend. In most of the regions, land degradation is very serious. Ecosystem integrity and biodiversity are seriously threatened. Other problems include exacerbated water shortage and quality degradation, marine pollution, coastal zone degeneration, and increased

[10]See Footnote 8.

[11]http://www.adb.org/sites/default/files/publication/160067/ewp-432.pdf.

[12]http://www.adb.org/sites/default/files/publication/160067/ewp-432.pdf.

[13]UN: 2012.

urban solid wastes. The mixture of existing problems and emerging issues adds difficulties in finding proper solutions.

In addition to the above-mentioned issues, the environment situation of China and ASEAN will display different features in the future 5–10 years.

#### 6.2.2.1 Overall Situation and Trends for Environmental Protection in China

The implementation of National 12th Five-Year Plan for Environmental Protection has guaranteed positive progress in environmental protection when national economy was advancing fairly fast. Total emission of major pollutants decreased, and deterioration of ecological environment was controlled. During the 13th Five-Year Period (FYP), China still faces severe air, water, and soil pollution. Emissions from production and daily life, urban and rural areas, industrial and transportation sectors would interweave with each other, and environmental problems become more complicated due to mixed pollution problems. With the fast advancement of industrialization and urbanization, as well as the shift from traditional growth engines to the new ones, environmental protection tasks remain challenging and greening development mode still has a long way to go.

During the 12th FYP, the Chinese government has given environmental protection an unprecedentedly high priority in its agenda. Ecological civilization was integrated into the overall national strategy of China to form a Five-in-One Layout for development together with economic, political, cultural, and social development in the top-level design. In 2015, Beijing released the Opinions on Accelerating the Development of an Ecological Civilization, which for the first time proposes the coordinated advancement of new-type industrialization, informationization, urbanization, agricultural modernization, and greennization to boost green, circular and low-carbon development for building a beautiful China. The 12th FYP also witnessed the formal enforcement of new Environmental Protection Law, as well as promising legislative progress for air, water, and soil pollution prevention and control. Action Plan for the Prevention and Control of Air Pollution, Action Plan for the Prevention and Control of Water Pollution, Action Plan for the Prevention and Control of Soil Pollution were issued one after another. Driven by the multiple policy measures, during the 13th FYP, pollution emissions are expected to reach the peak and decline by 2020.[14] Resource and energy consumption would stay at a high level, yet their increase speed will slow down. The new-type urbanization which emphasizes optimized layout and resource/energy efficiency will create synergies with green and low-carbon economic growth. The Belt and Road Initiative, Jing-Jin-Ji Economic Integration Strategy, and Yangtze River Economic Belt Plan will provide golden opportunities for regional and international cooperation on environment and development.

---

[14]http://news.xinhuanet.com/energy/2015-01/29/c_127435090.htm.

The 13th FYP defined the target of achieving an overall improvement of the ecological environment, i.e., strengthening inputs to eco-environmental protection to address the outstanding ecological and environmental issues, increase resource utilization efficiency and provide more and better ecological goods and services for people. Chinese government will fasten the pace of eco-environmental governance reform and modernize governance capability. Green and environmental protection industry will also embrace greater development potential. They will not only develop into the new areas for economic growth, but also help to increase efficiency, save resources, and provide high quality environmental goods.

#### 6.2.2.2 Overall Situation and Trends for Environmental Protection in ASEAN

As most of ASEAN economies are in the process of rapid urbanization and industrialization, in which manufacture serves as the pillar for economic growth, their industrial clusters are built more on the relative advantages brought by low costs rather than competitiveness from innovations. Therefore, development is imposing great pressures on environment. In light of fast economic globalization, the region will see and experience the growing weight of overexploitation of natural resources and environmental pollution.

The major environmental problems for ASEAN include: forest coverage loss, water and air pollution, biodiversity loss, etc. Statistics say the forest coverage in ASEAN reaches 43%, which makes the region among areas with the highest forest coverage in the world. However, growing population, expanding agricultural and industrial lands, deforestation, and mining are taking their toll on the region's forests. ASEAN countries had lost a total of 555,587 $km^2$ of forests between 1980 and 2007.[15] During 2000 and 2007, the forest area shrank from 2.09 to 1.9 million $km^2$, which equaled to an annual deforestation rate of 1.11%, much higher than the world average rate of 0.16%.[16] And the mangrove coverage in 2005 has declined by 26% compared with the 1980 level.[17] As of water and air quality, on the one hand, water pollution does exist in ASEAN region, on the other hand, water demand in the region is growing. If there are no adequately prompt and decisive measures, some places in the region would probably suffer from water shortage by 2030.[18] Some ASEAN Member States are experiencing bad air quality. In addition to industrial emissions, transportation sector is becoming a major source of emissions to affect air quality in the region. At subregional level, haze is a most significant transboundary environmental problem. Moreover, as a result of climate change, alien species invasion, wildlife poaching, and smuggling, species loss is in a critical

[15]http://env.asean.org/?p=640.

[16]Fourth ASEAN State of the Environment Report 2009.

[17]See Footnote 15.

[18]Asian Development Bank: ASEAN 2030: Toward Borderless Economic Community.

situation. According to IUCN Red List of Threatened Species, a large number of species in ASEAN are threatened and endangered, in which 1092 species are in Malaysia, 976 in Indonesia, and 944 in the Philippines.[19]

ADB studies show that lack of long-term effective management of environmental problems would likely threaten ASEAN economic growth. Fresh water shortage will constrain industrial and agricultural expansion, and substantially reduce life quality. Air pollution will cause higher health costs and production costs in many sectors. Environmental damages will also lay additional burdens on the poor population who are usually victims of unsafe water and natural disasters.

In the face of challenging environmental problems, economic growth and improvement of environmental awareness enable environmental governance to release its potential. Regional environmental initiatives and mechanisms will help resolve regional environmental problems. Since the release of the first ASEAN Subregional Environment Plan (ASEP-I), environment has always been a major item on ASEAN's development agenda. In 1981, the ASEAN Ministerial Meeting on the Environment (AMME) was established. ASEAN Member States have made positive strides in conserving land resources, setting up marine protected areas, managing air quality, addressing climate change, and other important cooperative areas for environmental protection and sustainable development.[20] The ASEAN Socio-Cultural Community Blueprint (2009–2015) has envisaged "ensuring environmental sustainability" as one of its characteristics and identified 11 areas [21] with strategic objectives and actions under the framework. It is acknowledged that positive policy messages will play a critical role in finding solutions to environmental problems in ASEAN.

### 6.2.3 *Environment and Development Opportunities for China and ASEAN*

In the future, China-ASEAN region will embrace a series of development trends which may have huge impact on ecology and environment and resource utilization, including:

---

[19]Fourth ASEAN State of the Environment Report 2009.

[20]Pengbin, Liu Xiaoxue, Yang Zhenzhong: Resource & Environment Situation and Cooperation Potential in ASEAN.

[21]These areas include addressing global environmental issues, managing and preventing transboundary environmental pollution, promoting sustainable development through environmental education and public participation, promoting environmentally sound technology (EST), promoting quality living standards in ASEAN cities/urban areas, harmonizing environmental policies and databases, promoting the sustainable use of coastal and marine environment, promoting sustainable management of natural resources and biodiversity, promoting the sustainability of freshwater resources, responding to climate change and addressing its impacts, and promoting sustainable forest management (SFM). see: http://www.asean.org/wp-content/uploads/images/archive/5187-19.pdf.

Scientific and technological innovation and sharing economy will become important pillars for green development. Technological innovation in areas of Internet+, e-commerce platform, and big data facilitates the development of sharing economy. The sharing economy model for transportation, housing, and infrastructure will dramatically push up energy utilization efficiency, reduce the pressure on environment and shape a more environment-friendly lifestyle.

Integration of regional economy will further boost trade facilitation and help shape green industrial chain and value chain. Regional integration will enable more convenient exchanges of people and logistics. To promote the industrial system of green supply chain and shape green value chain will increase energy efficiency and reduce the negative impact on environment.

The Belt and Road Initiative and Asian Infrastructure Investment Bank as advocated by China will dramatically boost infrastructure construction and mitigate corresponding bottleneck in the region. Belt and Road Initiative, together with ASEAN Community, will provide driving force and capital support for green development in the region. Green finance will channel the capital to green investment and the new policy measures of green credit and green debt will help to circumvent environment risks.

The implementation of 2030 Agenda for Sustainable Development points out strategic direction and serves as the driving force for regional green development and South–South cooperation on environment. The 2030 Agenda for Sustainable Development marks the innovation of global environment governance model, with more comprehensive and systematic environmental targets covering air, water, ocean, biodiversity conservation, etc. Its monitoring system is full-fledged and binding, and quantitative indicators are defined.

## 6.3 Establishing a Community for ASEAN-China Environmental Cooperation: Toward Green Development Together

### *6.3.1 Progress of ASEAN-China Cooperation on Green Development*

Environmental protection is among the priority areas for ASEAN-China cooperation, which lays solid foundation for further regional cooperation on green development. As China and ASEAN share the common goal of establishing resource-efficient and environmentally friendly societies, actions for protecting environment, reducing pollution, and checking ecological degradation will benefit the two sides, so does the enhanced cooperation in these fields. In 2003, leaders from China and ASEAN Member States adopted the Joint Declaration on Strategic Partnership for Peace and Prosperity to "further activate exchanges in science, environment, education, culture, personnel exchange … and improve cooperation mechanisms in these areas." In November 2007, the Chinese leader put forward a proposal at the 11th China-ASEAN Summit in Singapore to set up a China-ASEAN environmental protection cooperation center and develop a strategy for ASEAN-China cooperation. In 2009, China and ASEAN approved the China-ASEAN Strategy on Environmental Protection Cooperation (2009–2015), which identifies cooperative objectives, principles, and six priority areas for action. In 2010, the Ministry of Environmental Protection of China launched the China-ASEAN Environmental Cooperation Center. ASEAN and China adopted in 2011 the ASEAN-China Environmental Cooperation Action Plan (2011–2013), in 2013 the Action Plan of China-ASEAN Environmental Cooperation (2014–2015), and in 2016 the ASEAN-China Strategy on Environmental Cooperation (2016–2020), respectively. The succeeding ASEAN-China Environmental Cooperation Action Plan (2016–2020) has been drafted as well with concerted efforts from both sides.

Under the framework outlined by the ASEAN-China cooperation strategy and action plans, the two sides have carried out numerous practical actions in terms and areas of high-level policy dialogue, China-ASEAN Green Envoys Program, biodiversity and ecological conservation, environmental industry and technological exchanges, as well as joint research. The two sides organize a China-ASEAN Environmental Cooperation Forum every year since 2011 to discuss environmental issues of common concerns, which functions as a platform for China and ASEAN Member States to conduct high-level policy dialogue and facilitate exchanges and feasible actions. In October 2011, the China-ASEAN Green Envoys Program was launched, with the purpose to raise public awareness of environmental protection and facilitate environmental education. This program consists of green innovation (i.e. capacity-building activities for decision makers in the government), green pioneer (i.e., publicity and visibility campaigns for youth), and green entrepreneurs (i.e., building regional partnership on green economy in business sectors). Such multi-faucet program has greatly enriched the China-ASEAN environmental cooperation. With the support from Chinese Ministry of Environmental Protection (MEP) and ASEAN Secretariat, China-ASEAN Environmental Cooperation Center (CAEC) worked closely with ASEAN Center for Biodiversity (ACB) and developed a "China-ASEAN Cooperation Plan for Biodiversity and Ecological Conservation," which aims to facilitate design and execution of policies, plans and actions for biodiversity and ecological conservation. In consultation with MEP and the ASEAN Secretariat, CAEC also developed an ASEAN-China Cooperative Framework for Environmental Technologies and Industries, to depict a concrete road map for the bilateral cooperation in related fields.

### *6.3.2 Priority Areas for China-ASEAN Cooperation on Green Development*

China and ASEAN share the common goal of realizing the 2030 Agenda for Sustainable Development. As the core of the 2030 Agenda, SDGs emphasize the close link between human well-being, economic prosperity, and healthy environment, creating a favorable condition for China-ASEAN cooperation on green development. It is necessary to consider the future trend of environment and development, SDGs implementation requirements, as well as the environmental concerns and development needs of China and ASEAN Member States for better bilateral cooperation on green development in the future.

With the fast economic development and intense exploitation of natural resources, significant changes took place in regional ecology and environment, and a series of environmental issues grew prominent. These changes not only posed a severe challenge to the long-term social stability and economic development, but also fundamentally caused damages to the nature and ecosystem on which human beings depend for survival. As China and ASEAN Member States have different levels of socioeconomic development and environmental protection, both sides have different priorities in environmental sector. However, both sides confront common challenges in water pollution, air pollution, and biodiversity loss. With the improvement in the recognition of the significance of environmental sustainability, both sides share growing concern and inputs into green development, and there are growing needs in international cooperation. Meanwhile, priority areas in green development, such as sustainable production and consumption (SCP) and green economy, increasingly draw attention from the world.

Currently, the cooperation between ASEAN Member States and China in the field of environment and development is mainly based on China-ASEAN Environmental Cooperation Strategy and China-ASEAN Action Plan on Environmental Cooperation. Based on the successful implementation of China-ASEAN Strategy on Environmental Protection Cooperation (2009–2015), China and ASEAN updated a version of ASEAN-China Strategy on Environmental Cooperation (2016–2020). Its overall objective is to strengthen ASEAN-China cooperation in priority areas of environmental protection by taking a coordinated and integrated approach with a view of achieving environmental sustainability in the region. There are five specific objectives: (a) enhancing high-level policy dialogue with focus on environmental issues of common concern to increase understanding, enhance cooperation, and secure the harmonization of interests of ASEAN and China; (b) enhancing dialogue and cooperation on environmental protection; (c) improving capacity for national and regional environmental management through sharing knowledge and experiences and implementing joint actions; (d) enhancing cooperation on priority areas, improving effectiveness and quality of cooperation, and developing good practices for regional and South–South[22] environmental cooperation; and (e) supporting ASEAN Community's Post-2015 Vision. Considering the priorities set out by the new strategy and the situation of green development in China and ASEAN, both sides need to actively explore new paths in cooperation and facilitate cooperation in the nine priority areas as below.

[22]South–South Cooperation is a term generally used to describe the exchange of resources, technology and knowledge between developing countries.

#### 6.3.2.1 Policy Communication and Dialogue

Policy communication and dialogue will provide platforms for environmental policy makers in China and ASEAN to exchange views on key regional environmental issues, share the progress and achievements on ecological and environmental protection, and improve cooperation through joint efforts. It will help implement the consensus reached by leaders of China and ASEAN, forge a more comprehensive regional environmental cooperation system, and explore effective pathways for close and coordinated development in the region with environmental protection as the window.

#### 6.3.2.2 Environmental Data and Information Sharing

It is necessary to enhance China and ASEAN's capacity in collecting, processing, and utilizing environment data and information, including to establish China-ASEAN environmental information sharing platform as proposed at 17th China-ASEAN Summit; share the knowledge and experience on environment information and data, and conduct capacity-building activities on the collection, processing and utilization of environmental information and data.

#### 6.3.2.3 Environmental Management Experience Sharing

It is advised to conduct all round exchanges and cooperation on the policy, system, tools, and methods of environment management concerning the common issues in the area of environment and development, exchange the success stories, learn from each other the effective management systems and means, conduct joint studies on eco-environmental planning and management, and improve environment management capability and efficiency of China and ASEAN by sharing knowledge and experience.

#### 6.3.2.4 Biodiversity and Ecological Conservation

It is necessary for the Chinese side to cooperate with ASEAN Center for Biodiversity to further develop and implement China-ASEAN Cooperation Plan on Biodiversity and Ecological Conservation; improve China and ASEAN Member States' capacity and awareness in developing policies, strategies and/or action plans pertinent to biodiversity conservation; honor Convention on Biological Diversity (CBD) and other international obligations; promote the conservation, management and sustainable utilization of biological resources; and make good use of the economics of ecosystems and biodiversity.

#### 6.3.2.5 Facilitating Environmental Industry and Technologies for Green Development

It is important to implement the China-ASEAN Cooperation Framework on Environmentally Sound Technologies and Industries by establishing an information exchange platform, conducting demonstration projects and joint researches on the development of environmentally sound technologies, and participate in and support UN's 10-Year Framework of Programmes on Sustainable Consumption and Production. It is necessary to establish exchange and cooperation bases for environmentally sound technologies in appropriate localities, forge the bridge and pillar for the cooperation on environmental industries and technologies and promote the exchanges and cooperation of environment protection enterprises, environment governance technologies, and investment and finance in this area.

#### 6.3.2.6 Environmentally Sustainable Cities

It is useful to share the practice and experience on sustainable city development by establishing city-level cooperation network and partnership, and improve the capability of China and ASEAN Member States in sustainable city planning and construction, in particular the capability of small and new city regions.

#### 6.3.2.7 Environmental Education and Public Awareness

It is important to support the implementation of ASEAN Environmental Education Action Plan (2014–2018) and enhance public awareness in China and ASEAN Member States of environment protection via the exchanges and cooperation among environment education agencies, relevant government departments and civil society organizations between the two sides.

#### 6.3.2.8 Institutional and Personnel Capacity Building

It will generate fruitful results to increase the capability of China and ASEAN Member States on environment management via various capacity-building activities under the Green Envoy Program, including to strengthen the training for environment managerial staff, improve the capability of policy making in the areas of environmental economics, environment, and health; establish experience sharing platform for environment protection laws and regulation; and conduct two-way visits and exchanges for environment management staff to improve their capability.

#### 6.3.2.9 Joint Research

It is advised to promote communication and exchanges among scholars, think tanks, and research institutions to establish green think tank for China and ASEAN Member States; study global and regional emerging issues related to environment and development interested by both China and ASEAN; share the research results via the existing China-ASEAN cooperative mechanism; and provide targeted, scientific and information-based policy recommendations for policy makers.

### *6.3.3 Priority Actions for China-ASEAN Cooperation on Green Development*

It is of great significance to make top-level planning and specific projects design; actively take concrete actions on high-level policy dialogue, joint research, capacity building, industrial cooperation, demonstration projects and so forth; and further promote pragmatic cooperation between ASEAN and China in the nine priority areas of green development. To improve the effectiveness and quality of cooperation will benefit the implementation of Belt and Road Initiative and ASEAN Community, achieve environmental sustainability in this region, and provide best practices for South–South environmental cooperation.

#### 6.3.3.1 Carrying Out Policy Dialogue and Exchange, Forging Partnership, and Expanding Cooperation Mechanisms

As it is important to materialize the consensus reached by leaders of China and ASEAN via the existing high-level forums and ministerial conferences, the first suggestion is to update current China-ASEAN Environmental Cooperation Forum to China-ASEAN Environment and Development Cooperation Forum, expanding the scale and participation from government departments, businesses, research institutions, and NGOs. The upgraded forum will provide a platform for experience sharing in green development between governments, private sector, academic circle, and the public of China and ASEAN Member States, as well as facilitate discussion on building a green Maritime Silk Road.

It is recommended to actively build the partnership between government, business, and the public and establish a special cooperation mechanism. It is necessary to prioritize the partnership between national and local governments and enterprises of both sides in water pollution treatment, facilitating policy dialogue and experience sharing. It is necessary to establish the China-ASEAN cooperation platform in air pollution treatment, promoting communication and exchange with regard to air policies, enforcement, and energy conservation. Dialogues are suggested in the field of SCP and green economy between China and ASEAN Member States and explore the possibility of cooperation and demonstration projects.

Coordination and cooperation should be enhanced between the government, business, think tank, and international organizations on environment and development in China and ASEAN Member States, starting from strengthening dialogue and communication.

#### 6.3.3.2 Environment Data and Information Sharing

It is recommended to establish information sharing platforms and databases on environment and development between China and ASEAN, with joint research achievements and information related to environment and development of China and ASEAN Member States updated into the database. Information services will be provided via Web sites.

#### 6.3.3.3 Biodiversity and Ecological Conservation

China and ASEAN need to further strengthen domestic efforts and international cooperation on biodiversity conservation, improve respective capability and awareness in developing policies, strategies or action plans on biodiversity conservation and sustainable utilization; fulfill responsibilities set out by Convention on Biological Diversity, its protocols and other international obligations; push regional cooperation on biodiversity and ecological conservation; and promote the protection, management and sustainable utilization of regional biological resources.

It is necessary to share experiences in urban and rural ecological protection and jointly carry out demonstration projects; strengthen the capability in biodiversity protection, to create synergies with poverty reduction and climate change adaptation; promote the cooperation in priority areas, e.g., the cooperation on ASEAN relics parks, biodiversity monitoring, etc.; and research on biodiversity and ecological policy tools and practice.

Research efforts should be made to quantify the economic value of biodiversity, in particular the evaluation of ecosystem services using various economic models; and promote application of the economics of ecosystems and biodiversity (TEEB) in China and ASEAN Member States to inform their green transition.

#### 6.3.3.4 Promoting Environmental Technologies and Industries to Support Green Development

Recommended activities include: (a) promote cooperation on environmental industries and exchange of environmental technologies on the foundation of efforts paid to policy dialogue and capacity building; (b) implement China-ASEAN

Cooperation Framework on Environmentally Sound Technologies and Industries and support Ten-Year Framework on Sustainable Consumption and Production; and (c) establish a China-ASEAN Cooperation Network on Environmental Technologies and Industries, inviting the wide participation from relevant government departments and the industry.

The Cooperation Network is divided into two layers. To begin with, it includes a long-term communication mechanism, through establishing focal points from the environment-related departments of national governments, so as to consolidate exchange and information sharing with regard to environmental industry planning and policies. Secondly, it includes a mechanism of information exchange and dialogue for industry associations and chambers of commerce in China and ASEAN.

It is recommended to establish China-ASEAN Platform of Information Exchange in Environmental Technologies and Industries, releasing policies on environmental industries, news, standards and regulations, and product description, etc., of all countries through portals and Web sites. Provision of information services should be given particular emphasis.

It is necessary to forge a platform of R&D and training to bolster strategic cooperation between research institutions and business in the field of environment protection and development. Specifically, the platform can serve to organize R&D and training, to address problems faced by the industry, improve technologies of enterprises of both sides, and underpin environmental governance capacity.

It is also advised to establish a cohort of China-ASEAN demonstrative bases for international cooperation and exchanges of environmentally sound technologies and industries based on the existing bases in Yixing, Jiangshu Province and Wuzhou in Guangxi Zhuang Autonomous Region for the display of environmental industry and technologies of China and ASEAN Member States. These bases will support scientific research, education and training, and industrial cooperation and become a coordinated innovative platform for the integration of industries, universities, and research institutes. Efforts are needed to promote the joint R&D of ecologically sound technologies, and establish transfer markets to facilitate the development and transfer of environmentally sound technologies.

#### 6.3.3.5 Joining Hands to Develop Environment-Friendly Cities

Programs should be designed to increase the capability of China and ASEAN Member States in promoting the construction of environmentally sustainable cities; share the knowledge and experience in urban ecological protection, establish expert cooperation network to promote ecologically friendly development of urban areas; strengthen the cooperation on sustainable production and consumption in the context of urbanization to help shape the industrial structure, growth approach and consumption model featuring resource conservation and environment protection; and cooperate on non-hazard treatment of urban wastes.

#### 6.3.3.6 Advancing Personnel Exchange and Capacity Building

To realize regional green development, it is necessary to ensure that macroscopic regulation on socioeconomic activities must be undertaken within the environmental carrying capacity, decoupling economic growth from resource consumption and environment degradation. Therefore, improvements are needed with regard to the capacity in environmental governance and green development of China and ASEAN Member States.

Related capacity-building fields and activities may include: (a) enhancing the capacity of environmental officials with regard to policy formulation and enforcement in green economy, environment economy, environment and health, etc., through training on comprehensive capacity; (b) enhancing capacity in environmental enforcement and facilitating information sharing on environmental policies and enforcement; (c) organizing mutual visit and personnel exchange between China and ASEAN Member States in the management level, improving the comprehensive capacity in environmental governance; (d) strengthening training and exchange in EIA technologies; (e) enhancing capacity building in biodiversity conservation to facilitate the synergetic interaction between biodiversity conservation, poverty relief and response to climate change; and (f) carrying out exchange and training in waste recycling, and promoting relevant knowledge among government decision makers, the business and the public.

The "Green Envoys Program for the Maritime Silk Road" should be carried out based on existing China-ASEAN Green Envoys Program, facilitating cooperation in personnel exchange, joint training and policy dialogue to improve capacities in the field of green development.

It is necessary to enhance capacity building in green development education and training. It is suggested to (a) establish China-ASEAN Education Network on Green Development, organizing seminars at different levels periodically to share education experience in green development from all countries; (b) carry out capacity building and training for environmental education institutions in China and ASEAN Member States, to improve the capacity of training and education institutions in policy support, project implementation, financing and achievement promotion; (c) compile and publicize materials and brochures on green development; and (d) conduct joint research on environmental education, environmental information, public participation and other common concerns.

#### 6.3.3.7 Deepening Industrial Cooperation and Exchange of Technologies

It is recommended to promote the exchanges and capacity building of scholars and think tanks to forge the green think tanks of China and ASEAN. Efforts are needed to track and analyze sustainable development issues, hot environmental issues and key issues in the global and regional levels, delve into the prime concerns in the priority fields of green development and priority areas of cooperation, explore the development of innovative policy tools, and provide theories and knowledge, policy recommendations and decision-making tools for policy makers. It will be useful to build a network of experts and set up a list of topics for joint research, based on the priority areas for study.

The primary key issues for joint researches are: green transformation of industries, priority sequence in industrial transformation, green investment and trade, green infrastructure construction and resources and environment protection, resource efficiency improvement, recycling and remanufacturing and green finance, etc. Demonstration can be carried out based on country-specific situations.

The two sides can also have joint research and knowledge sharing for implementing SDGs and climate change actions.

#### 6.3.3.8 Carrying Out Pilot Programs to Prepare for Best Practice Dissemination

It is advised to carry out demonstrative projects in environmental technologies as well as environmental goods and services. In water pollution, air pollution and solid waste disposal and treatment, environmental monitoring and other fields of common concern by China and ASEAN Member States, the two sides will select applicable environmental technologies and carry out demonstrative pilot programs, which will help to discuss and address issues on technologies, standards, and operating models faced by the environmental industry of China and ASEAN Member States.

Other priority actions include:

Screening environment-friendly and clean production technologies: based on their situations and practical needs in green development, China and ASEAN are expected to identify the pollution monitoring technologies and equipment that are most suitable for local conditions, pollution treatment technologies and equipment, clean production products and technologies, and resource recycling technologies, to formulate and release the "Inventory of Technologies for China-ASEAN Cooperation on Environmental Industry."

Pilot programs on pollution treatment technologies: according to the results of technology screening, demonstrative programs should be developed to pilot particular technologies selected through special seminars. Experience from the implementation of the demonstrative programs will be used for the promotion and transfer of pollution treatment technologies in the future. Priority should be given to water treatment technologies and industry.

Cooperation and mutual recognition of environmental labeling (eco-labeling) products: it is necessary to cooperate on environmental labeling products, exploring ways to establish a common green or environmental labeling system, and facilitate China and ASEAN Member States to achieve agreement on mutual recognition of environmental labeling (eco-labeling) products, by means of holding seminars on technologies, policy consultation meeting and so on. It is also necessary to expand the market share of environmental labeling (eco-labeling) products, and explore patterns and measures to put related policies into practice and action.

Other priority demonstrative projects: it is recommended to institute technology transfer market in the domain of environmental industry to rev up the development and transfer of environmentally sound technologies and clean production technologies. Also, it is significant to carry out pilot and demonstrative programs in appropriate sectors on green supply chain based on existing research in this field. Thirdly, it is necessary to carry out cooperation and demonstrative programs on SCP in the context of urbanization, to expedite the formulation of the industry structure, growth pattern and consumption pattern featuring energy conservation and environmental protection. Fourthly, demonstration pilot programs should be developed for biodiversity and ecological monitoring.

#### 6.3.3.9 Promoting Cooperation on Sustainable Consumption and Production

Specific activities are recommended as follows: (a) further facilitate the dialogue between China and ASEAN on sustainable production and consumption; (b) conduct joint researches on key issues of national strategy and road map for sustainable consumption and production, policy effect evaluation, sustainable product innovation and market opportunities, status quo and development of regional environmental goods and services, green supply chain, green procurement, and green trade, and (c) carry out policy pilots on environmental labeling and clean production, with efforts to probe into models and practices to transfer policy orientation to real actions.

### *6.3.4 Outlook into the Future: The Environment and Development of China and ASEAN*

As stated in the previous chapters, in general, the environment of China and ASEAN Member States is severe and there is still a long way to go for realizing green development, which not only requires the implementation of effective national policies for all the countries, but also needs strengthening international and regional cooperation and joint efforts to address the challenges. Environmental issues emerged during the rapid economic development, population growth, and the accelerated industrialization and urbanization, such as pollution, the excessive exploitation of natural resources, loss of biodiversity, ecosystem degradation and so on have made China and the ASEAN Member States realize the necessity of green development. The two sides have carried out pragmatic and effective regional cooperation in the fields of environmental protection and development, and explored new paradigm for South–South cooperation, which achieved positive results and laid a solid foundation for creating a community of environmental cooperation.

To make the common vision toward green development come true, China and ASEAN Member States need to further improve their policy framework, continue to innovate policies and financing models both at the macrolevel and microlevel, transfer the existing national policies in development, energy and environment into concrete actions, and effectively cope with such challenges as unbalanced development, environmental pollution and ecological degradation, weak governance, inadequate funding, lack of technologies and innovation, etc.

Looking into the future, China and ASEAN Member States are expected to grow into key drivers for the economic development in Asia-Pacific region. It is believed that with concerted efforts and cooperation, China and ASEAN Member States can achieve green development with effective approaches, policy innovation and pragmatic actions. In the meanwhile, with the historical opportunity of the kick-off of 2030 Sustainable Development Agenda, China's Belt and Road Initiative and ASEAN Community, the two sides will continue to forge ahead in the exchanges and cooperation for green development and environment protection to jointly facilitate the realization of green, low-carbon and sustainable development goals

# Working Group on China-ASEAN Environment Outlook

**Chair and Leading Author: Dr. Zhou Guomei** has more than 20 years experience as an environmental economist and policy researcher for a number of studies on environmental and economic policies, providing research support for national decision—making for environmental protection in China. Dr. Zhou was senior visiting scholar of MIT of USA, Kyoto University of Japan and York University of the UK. She also worked as consultant and environmental economist expert in a number of studies for UNEP, World Bank, ADB, JICA, and other international sponsors. Dr. Zhou Guomei is the state candidate of New Century National Hundred, Thousand and Ten Thousand Talent Project in 2009, which issued by the Ministry of Human Resource and Social Security and other seven relevant ministries of China. She was granted in 2012 Special Government Allowance from the State Council. Her academic interests are global environmental issues, international environmental policies, environmental economics, economic evaluation of environmental impacts, etc. She has published more than 100 research papers and 12 books (including co-authors). Dr. Zhou Guomei is Deputy Director General of China-ASEAN Environmental Cooperation Center of Ministry of Environmental Protection of China, and Deputy Coordinator of Chief Advisor Support Team of the China Council for International Cooperation on Environment and Development (CCICED).
E-mail: zhou.guomei@chinaaseanenv.org



China-ASEAN Environmental Cooperation Center, *China-ASEAN Environment Outlook 1 (CAEO-1)*, https://doi.org/10.1007/978-981-10-6211-7

**Core Member and Leading Author: Dr. Shi Feng** got his Ph.D. from the University of Georgia with a major of resource and environment research. After graduation, he worked in environmental protection governmental departments in both the USA and China. His recent academic interests include environmental laws and regulations, pollution prevention and control, water environmental protection, international cooperation on environmental protection, regional development strategy. He has published a number of articles and research papers, in Chinese and English and three books (including co-authors). Dr. Shi Feng is now the Deputy Director of China-ASEAN Environmental Cooperation Center of Ministry of Environmental Protection of China.
E-mail: shi.feng@chinaaseanenv.org

**Core Member and Leading Author: Jiang Hongqiang** Ph.D. and Professor, graduated from Renmin University of China, majored in Resources and Environment Economics. His present research focus lies in Environmental Planning and Policy. Now he is the Executive Director of National Key Laboratory of Environmental Planning and Policy Simulation, Chinese Academy for Environmental Planning (CAEP). He is also a consulting expert of several international projects funded by the World Bank, Asian Development Bank, and projects such as China-EU project, China-Sweden project, China-India project. He has been chairing or participating over sixty researches sponsored by the National Natural Science Foundation, National Philosophy and Social Science Fund, National Science and Technology Major Project, National Key Technology R&D Program up to now. With eight works published, and other eight anticipating. Publish over ninety papers. He has won eight Provincial Science and Technology Awards. Now he is holding a concurrent post as Member and Deputy Secretary General of the Environmental Planning Committee under Chinese Society for Environmental Sciences, Member of Environmental Economics Branch of Chinese Society for Environmental Sciences, Council Member of Environmental Auditing Branch under Accounting Society of China, and Editor of Environmental Decision-Making Reference, and Chinese Journal of Environmental Management.
E-mail: jianghq@caep.org.cn

**Core Member and Leading Author: Su Yang** holds a Ph.D. of Environmental Sciences and served in the Development Research Center of the State Council since 2002. His main research area is policy analysis on environment and conservation of cultural and natural heritage. He has led and conducted a number of research projects at the level of state and ministries, including the Financial Mechanism of Nature Reserve in China, the Quantitative Appraisal of Cultural Heritage Management System, Environment and Health Sub-Project of the Macro Research on National Environment Strategy, Environment and Health Governance Reform Project, Research on Main Functional Area-related Population Strategy, etc. He has won the Grand, First and Second Prize of the "China Development Research Award" and published six books in the area of protected area as the main author.
E-mail: suyang1@drc.gov.cn

**Core Member and Leading Author: Ge Chazhong** Professor and Director of the Institute of Environmental Policy, Chinese Academy of Environmental Planning, vice chairman/general secretary of Chinese Society for Environmental Economics of the Chinese Society of Environmental Sciences, has been working on environmental policy studies, environmental planning, the use of economic instruments for environmental protection and environmental financing since 1997. He has currently been involved in the projects such as the draft of 13th national environmental protection planning, pilot implementation of national environmental economic policies, national environmental taxation policy design and payable obtain of pollutants permit and emission trading. As environmental policy specialist, he has been involved in many international projects such as the work of UNEP including the GEO-4–GEO-6 and IWQGES.
E-mail: gecz@caep.org.cn

**Core Member and Leading Author: Huang Haifeng** obtained his Ph.D. at Humboldt University Berlin and studied at the Johns Hopkins University-Nanjing University Center for Chinese and American Studies and the School of Economics of Peking University. Professor Huang is the Director of Center for Green Economy at Peking University HSBC Business School. Professor Huang is serving as international jury of Sustainable Entrepreneurship Award in Europe and a member of the supervisory committee of the Ministry of Land and Resources. He has also published more than 10 books on sustainable development. In 2015, Professor Huang received the Outstanding Innovation Award of the Education Forum for Asia, as well as the Lifetime Award from the 3rd International Conference CSR, Sustainability, Ethics and Governance (Cologne).
E-mail: huanghf@phbs.pku.edu.cn

**Core Member and Contributing Author: Zhang Jinhua** works in the United Nations Environment Programme's Regional Office for Asia and the Pacific, as the Regional Sub-programme Coordinator on Environment Under Review. He plays a leading role in organizing the Asia-Pacific regional inputs to Global Environment Outlook reports (GEO-4, GEO-5, and GEO-6), as well as in guiding more than 20 countries and cities in development of state of environment reports. Jinhua has coordinated sub-regional and thematic scientific assessments including The Pacific Environment and Climate Change Outlook, South Asia Environment Outlook, water vulnerability assessments in Northeast Asia, South Asia, and Southeast Asia. He has been serving as members of advisory group for global forest assessment 2005 and SUMERNET project of Stockholm Environment Institute, respectively. He has led several partnerships on integrated environmental assessment, environmental data, and information sharing platform. Jinhua has degrees in Science and Agronomy from Wageningen University, The Netherland and Nanjing Agricultural University, China.
E-mail: jinhua.zhang@unep.org

**Core Member and Contributing Author: Lailan Syaufina** is a Lecturer of Faculty of Forestry and Researcher of Study Center for Disaster Management, Bogor Agricultural University, Indonesia. Her expertise includes sustainable tropical forest management, forest health monitoring, peat fire management, forest fire prevention and control, as well as environmental damage assessment. She is the author of a number of scientific publications and a member of several scientific panels and research programs. She received her Ph.D. degree from Faculty of Forestry University Putra Malaysia, Malaysia, and M.Sc. Forest. Trop. degree from Faculty of Forestry of Georg-August University of Goettingen, Germany.
E-mail: syaufinalailan@gmail.com; lailans@ipb.ac.id

**Coordinator: Peng Ning** graduated from Tsinghua University with a MA degree in English literature. She is an engineer and program officer in China-ASEAN Environmental Cooperation Center of Ministry of Environmental Protection of China and assistant to Chief Advisor Support Team of the China Council for International Cooperation on Environment and Development (CCICED). Her recent academic interests focus on international environment and development theories, international environmental policies, regional mechanisms for environmental protection, and strategic policy study. She has published more than 10 papers and two books (including co-authors).
E-mail: peng.ning@chinaaseanenv.org

**Senior Advisory Group**

Xu Qinghua
Li Haisheng
Guo Jing
Song Xiaozhi
Tu Ruihe
Tan Qingrui
Zhang Jieqing

**Working Group on China-ASEAN Environment Outlook**:

**Chair**:

Zhou Guomei (China-ASEAN Environmental Cooperation Center, China).

**Core Members**:

Zhou Guomei (China-ASEAN Environmental Cooperation Center, China; Leading author for Summary Report for Decision Makers, Chap. 6, and report outline, and responsible for final compilation and editing)

Shi Feng (China-ASEAN Environmental Cooperation Center, China; Leading author for Chap. 6 and Summary Report for Decision Makers, and responsible for final compilation and editing)

Jiang Hongqiang (Chinese Academy for Environmental Planning, China; Leading author for Chaps. 1 and 2)

Su Yang (Development Research Center of the State Council, China; Leading author for Chap. 3)

Ge Chazhong (Chinese Academy for Environmental Planning, China; Leading author for Chap. 4)

Huang Haifeng (Peking University HSBC Business School, China; Leading author for Chap. 5)

Lailan Syaufina (Bogor Agricultural University, Indonesia; Contributing author)

Zhang Jinhua (UNEP; Contributing author)

**ASEAN Secretariat**:

Raman Letchumanan (Former Head of Environment Division of ASEAN Secretariat), Natalia Derodofa, Rocky Pairunan.

**Coordinator**:

Peng Ning (China-ASEAN Environmental Cooperation Center, China; Contributing to the Summary Report for Decision Makers, Chap. 6, report outline, and responsible for final compilation and editing, and project management and coordination).

**Contributing Authors**:

Lailan Syaufina (Bogor Agricultural University, Indonesia), Zhang Jinhua (UNEP), Zhang Jing (Chinese Academy for Environmental Planning, China; co-author for Chaps. 1 and 2), Wang Yufei (Development Research Center of the State Council, China; co-author for Chap. 3), Cheng Cuiyun (Chinese Academy for Environmental Planning, China; co-author for Chap. 4), Duan Yunting (Chinese Academy for Environmental Planning, China; co-author for Chap. 4), Tian Chaoyang (Chinese Academy for Environmental Planning, China; co-author for Chap. 4), Wu Huanan (Peking University, China; co-author for Chap. 5).

**Peer Reviewers**:

Li Lailai (World Resources Institute), Niramon Sutummakid (Thammasat University, Thailand), Clarissa Arida (ASEAN Center for Centre, the Philippines), Sheila Vergara (ASEAN Center for Biodiversity, the Philippines).

**Supporting Experts**:

Li Xia (China-ASEAN Environmental Cooperation Center, China), Chen Gang (China-ASEAN Environmental Cooperation Center, China), Shi Feng (China-ASEAN Environmental Cooperation Center, China), Peng Ning (China-ASEAN Environmental Cooperation Center, China), Shang Huijun (China-ASEAN Environmental Cooperation Center, China), Lan Yan (China-ASEAN Environmental Cooperation Center, China), Xie Ran (China-ASEAN Environmental Cooperation Center, China), Lv Xiani (China-ASEAN Environmental Cooperation Center, China), Tian Yuan (Ministry of Finance, China), Liu Mi (Peking University, China), Xu Sihan (Peking University, China), Zhang Wenqi (Peking University, China), Deng Chunfang (Peking University, China), Wang Chuansheng (Peking University, China), Wang Hui (Peking University, China).

**Expert Reviewers and Participants of Previous Meetings for China-ASEAN Environment Outlook**:

Guo Jing (Director General of China-ASEAN Environmental Cooperation Center, China), Song Xiaozhi (Deputy Director General of Department of International Cooperation of Ministry of Environmental Protection, China), Fang Bing (Division Director, Department of Asian Affairs of Ministry of Foreign Affairs, China), Sun Xuefeng (Director of Asian Affairs Division of Department of International Cooperation of Ministry of Environmental Protection, China), Chen Gang (Deputy Director of Asian Affairs Division of Department of International Cooperation of Ministry of Environmental Protection, China)

Raman Letchumanan (Former Head of Environment Division of ASEAN Secretariat),

Natalia Derodofa (ASEAN Secretariat), Rocky Pairunan (ASEAN Secretariat)

Jia Gensuo (Institute of Atmospheric Physics of CAS, China), Huang Yi (Peking University, China), Chen Ying (Research Center for Sustainable Development of CASS, China), Zhang Fengchun (Chinese Research Academy of Environmental Sciences, China), Ouyang Wei (Beijing Normal University, China), Zhang Kebin (Beijing Forestry University, China)

Jiang Nanqing (UNEP Beijing Office), Claudia Walther (GIZ), Dai Min (GIZ), Thuy Tien Phan Thi (GIZ), Wang Xiaojuan (Heinrich Böll Stiftung), Qin Hu (Environmental Defense Fund)

Zhang Jing (Chinese Academy for Environmental Planning, China), Tian Chaoyang (Chinese Academy for Environmental Planning, China), Han Linyan (Development Research Center of the State Council, China), Zheng Jun (China-ASEAN Environmental Cooperation Center, China), Li Panwen (China-ASEAN Environmental Cooperation Center, China), Huang Yiyan (China-ASEAN Environmental Cooperation Center, China), Hua Ruixiang (China-ASEAN Environmental Cooperation Center, China), Zhou Ye (China-ASEAN Environmental Cooperation Center, China), Zhu Bingeng (China-ASEAN Environmental Cooperation Center, China), Ge Lin (Beijing University of Technology, China), Ji Bingbin (Beijing University of Technology, China), Wang Xinyu (Beijing University of Technology, China), Du Jiemei (Beijing University of Technology, China), Ke Fei (Beijing University of Technology, China), Fang Wei (Tsinghua University, China), Zhao Huifen (China Environmental Science Press, China)

Sincere thanks should go to all those guiding, supporting, reviewing, and participating in project implementation and research!

MIX
Papier aus verantwortungsvollen Quellen
Paper from responsible sources
FSC® C105338

If you have any concerns about our products,
you can contact us on
**ProductSafety@springernature.com**

In case Publisher is established outside the EU,
the EU authorized representative is:
**Springer Nature Customer Service Center GmbH**
**Europaplatz 3, 69115 Heidelberg, Germany**

Printed by Libri Plureos GmbH
in Hamburg, Germany